B[illegible]

GREATER MANCHESTER

PETER B. HARDY

PGL Enterprises, Sale, Cheshire

Printed in Great Britain by Colin Cross Printers, 11 Leachfield Industrial Estate,
Green Lane West, Garstang, Preston, Lancashire, PR3 1PR.

Published by PGL Enterprises, 10 Dudley Road, Sale, Cheshire, M33 7BB.

1998

All figures and illustrations by the author unless otherwise stated.

ISBN 0 9532374 0 0

PHOTOGRAPHS

INSIDE FRONT COVER
Clouded Yellow *Colias croceus*, Mersey bank, Stretford 17.vii.92 (p.18)
Speckled Wood *Pararge aegeria*, Nan Nook Wood, Wythenshawe Park 13.viii.95 (p.40)
Gatekeeper *Pyronia tithonus*, near Stretford sewage farm 1.viii.96 (p.42)
Painted Lady caterpillar *Vanessa cardui*, New Manor Farm, Stretford (p.33)
Holly Blue *Celastrina argiolus*, garden in Sale 21.vii.92 (p.30)
Large White caterpillars *Pieris brassicae* on isolated crucifer in an alley in Sale 3.ix.92 (p.20)
Green Hairstreak *Callophrys rubi*, Dick Clough, Tunstead 14.v.94.
Brimstones *Gonepteryx rhamni*, reared indoors from caterpillars found in Sale water-park, 15.vii.94 (p.19)

INSIDE BACK COVER
Small Copper *Lycaena phlaeas*, var. *radiata*, Mersey Bank, Stretford 27.ix.89 (p.28)
Common Blue *Polyommatus icarus*, disused railway sidings near Horwich 26.vi.94 (p.29)
Small Heath *Coenonympha pamphilus*, moors east of Littleborough 25.vi.95 (p.44)
Narrow-Bordered Five-Spot Burnet *Zygaena lonicerae* and Six-Spot Burnet *Z.filipendulae*, Ardwick 25.vi.95 (p.45)
Camberwell Beauty *Nymphalis antiopa*, Waterside Hotel, Didsbury 2.viii.95 (p.36) (B.T.Shaw)
Pre-mating Small Tortoiseshells *Aglais urticae*, old nurseries, Sale/Northern Moor boundary 5.vii.95, site since destroyed (p.13)
Peacock *Inachis io*, Dumplington 14.viii.94, site since destroyed (p.13)

FOREWORD

Among the many and varied aspects of British wildlife, butterflies are especially treasured. They act as 'flagship' species for the vast number of poorly known insects that form a vital part in the functioning of natural systems. As such, they provide an essential link between the public and landscape; in particular, they can help to foster an appreciation of landscape health and how important this is to human health. In this way, butterflies are also indicator species. They are sensitive to change, changes in land use and management, to climate and pollution. Beautiful creatures of sunny summer days, easily recognised and observed, they can be readily mapped. The result, atlases such as this one for Manchester, combine with data for other regions and different organisms, to give a 'barometer' for British wildlife.

The present atlas of Manchester is, in several respects, an important contribution to the documentation of British butterflies. First, it is an atlas of those often-discarded regions, urban landscapes, frequently dismissed as wastelands for wildlife. This work clearly demonstrates that such a picture is far from true. There are sites throughout Manchester with a butterfly diversity as rich as that of rural areas beyond its boundary. If anything, it is the surrounding rural areas that are fast losing habitats and species from advances in modern agriculture. Second, this work is unique in the way the region is mapped at three different scales. The maps of part of the Mersey Valley at a resolution of 100 metre squares are an important innovation. They give a rare insight into urban/'rural' contrasts. They also highlight what is happening at local scales. Increasingly, it is becoming evident that habitats are being destroyed or damaged at scales too fine to be picked up by the resolution of mapping used in many conventional atlases. Third, it is also unique for providing detailed maps of butterfly habitats, hostplants and nectar sources. Combined with the fine scale of mapping it serves as a valuable tool, establishing a network of fixed quadrats for quantifying changes in habitats and their dependent species. The maps for urban cover in 1960 and 1996 show that urban development progresses unrelentingly. Future road developments are already planned for the Mersey Valley, across the area mapped in detail.

Altogether, this atlas represents an original concept and a remarkable achievement, a testimony to Peter Hardy's dedication and industry. For this, he deserves our gratitude and warmest congratulations. Peter himself has carried out much of the mapping, over two decades and during lunch times and weekends. It has clearly been a labour of love; the reader quickly becomes aware of an uncompromising and deep concern for wildlife and the environment in which we all have to live. It is a start, as Peter Hardy intends, that should provide much encouragement for the study of wildlife in those most difficult of landscapes, the city region.

R.L.H. Dennis
9:12:97

CONTENTS

PREFACE AND ACKNOWLEDGEMENTS.

The idea of an intensive multi-scale mapping project for the butterflies of the Manchester area was first suggested by R.L.H.Dennis in 1990. Many county butterfly atlases have been produced recently, but Manchester would not readily fall into any such, being outside the "modern" (post-1974) county of Cheshire (although many districts, including mine, Sale, still retain "Cheshire" in their postal address) and divided between two of the Watsonian vice-counties which are frequently used for biological recording (South Lancashire and Cheshire). Mapping the butterflies around a large city, rather than a county, was pioneered by Garland's (1981) atlas of butterflies of the Sheffield area. Garland adopted as his recording area a 50 X 30 km rectangle rather than an administrative unit. After due consideration the latter was however felt to be more suitable for the present work as a "rectangle" would have greatly increased the area to be covered, and as the sections outside the Greater Manchester boundary are far less accessible by public transport, a balanced coverage would have been difficult.

As the survey progressed, it became apparent that a small section of the area mapped in greater detail would provide much additional information on butterflies of urban and urban-fringe zones and their adaptability. Also, it would enable changes to urban biotopes in Manchester to be closely monitored. The mid-Mersey Valley was considered an appropriate sector for this as it contains a cross-section of habitats typical of much of the area.

I admit that having been born and bred a resident of "Cheshire" I decidedly did not welcome the new concept of a "metropolitan county" of "Greater Manchester", with its ten "districts" (fig.1) when it was introduced in 1974. After a few years, however, it became clear that there are in fact many advantages. The new boundaries have created a distinct geographical entity: a central city surrounded by urban and suburban districts which although formerly distinct towns in their own right now merge imperceptibly into one large whole, and are served by a radial network of road and rail routes (see fig.3). In the heyday of the 1980s virtually all parts were readily accessible in the orange-and-white buses of Greater Manchester Transport, and such bargains as "Sunday Rover" tickets and "Off-Peak Any-Distance Clippercards" made travel the length and breadth of the new self-contained "county" very economical. Although during the decade following bus "deregulation" this situation has changed for the worse, reasonably cheap travel by public transport within Greater Manchester is still sufficiently practicable for the area to form a convenient recording unit.

Here then is the result. I see it as only a beginning, and hope that it will arouse much more interest in Manchester's butterflies and stimulate more detailed work in the future. I also hope that it might increase the awareness of butterflies, including the common ones, as indicator species of the health of habitats, and perhaps a re-evaluation of what constitutes ideal land use in and around a city.

I would like to extend my gratitude to all those who have contributed to this survey:
J.Abbot, P.Alker, G.Allen, L.Almond, J.Anderton, K.Anderton, T.Appleton, C.Ashton, J.Ashworth, S.Bakht, H.Barlow, J.Barlow, N.E.Barnes, A.Barrett, A.Barton, R.Benham, G.Bennett, G.Bennion, D.P.Bentley, K.M.Berry, P.Berry, R.Biggs, Blackleach Country Park Wardens, C.Blythe, A.J.Bond, C.Boulton, S.Briggs, M.Z.Brooke, A.M.Broome, D.Broome, A.Brown, I.Brown, C.Buckley, B.Bull, J.Burgess, M.Burlinson, S.Burnet, I.Campbell, J.Chadwick, A.Clamp, K.Clayworth, W.Close, E.Cockshott, T.Coombs, S.P.Cotton, W.Cooper, D.E.Coupe, J.Courtman, G.Crossley, A.L.Crudge, A.J.Currie, J.Dagnall, C.Darbyshire, S.Darbyshire, K.Darwin, M.J.Davenport, A.Davison, I.& J.Davison, A.Day, E.Day, R.L.H.Dennis, J.Dowler, D.Downing, J.Dunn, D.P.Earl, E.Eckersley, P.Edmonds, W.Edwards, H.Entwistle, Etherow Country Park Wardens, J.Evans, D.R.Farnworth, I.Fitch, D.C.Fleet, D.Fleming, P.Foden, S.Forshaw, P.Fox, P.M.A.Francis, R.G.Gabb, B.M.Garland, S.P.Garland, P.Gateley, J.& M.Gibson, D.Godbert, H.Godfrey, S.Goodhew, E.Gorton, V.Grantham, R.Grayson, G.N.Green, S.Greenall, R.S.Greenwood, J.Grima, J.M.& D.C.Grindrod, J.P.Guest, J.Guthrie, B.Hancock, E.G.Hancock, P.A.Harper, T.Harper, S.Harrison, M.Harrop, G.C.Hardy, K.Haydock, S.J.Hayhow, J.Headon, I.Heap, P.R.Helm, M.Hemingway, T.Henshaw, M.Hickson, P.Hill (Rochdale), B.Hilton, H.& M.Hind, S.H.Hind, A.Holder, J.E.Holmes, R.Holmes, C.Hopley, M.Howard, L.E.Hughes, R.E.Hunter, J.Hurst, L.Hurst, G.Jackson, P.Jellen, M.F.Jenkinson, C.Johnson, C.Jones,

J.Jones, A.Kearsley, E.Kearns, P.Kearsley, H.M.Kelly, U.Kelly, G.R.Kenyon, I.Khalid, L.Kidd, P.Kidger, E.King, D.Knowles, B.Koo, J.Kubavak, B.Langridge, C.Lawlor, G.Lawson, A.B.Leach, G.Leather, S.Lee, H.Lees, M.& S.Lees, S.Leigh, G.Lightfoot, J.P.Lightfoot, M.Lightowler, P.Liptrot, M.G.Livingstone, P.Lord, W.Lowe, D.Lumb, K.McCabe, M.McCormick, A.McKeever, C.McLeod, M.McMillan, J.McNicholas, E.H.Makinson, H.Marsh, N.Martin, Medlock Valley Wardens, J.Mee, P.Menzies, Mersey Valley Wardens, A.& S.Middlehurst, F.Millet, D.Mitchell, Y.D.Mitchell, T.Morton, R.Murphy, M.Myers, R.Naik, K.R.C.Neal, New Mills Naturalists, C.Newton, D.H.Nightingale, P.North, M.Oates, Oldham Local Interest Museum, M.Orehtas, C.Owen, M.Parker, R.Parker, A.Partington, M.L.Passant, A.J.Percy, A.Phelps, D.G.Phelps, A.E.Pickford, D.Pickis, M.Pimblett, J.Platt, B.Porter, M.Prescott, I.Prior, K.L.Ransom, D.Reeves, M.Renouprez, R.Rhodes, J.Riley, A.Rimmer, H.Roberts, P.Robinson, D.Roscoe, P.Rowland, L.Rowley, D.Rugman, I.Ryding, St.Hugh's School (Oldham), S.J.Saunders, S.Schofield, B.T.Shaw, A.Shelley, L.Sivell, P.Slater, D.Smethurst, A.Smith, A.J.Smith, I.F.Smith, J.Smith, P.R.Smith, S.B.Smith, D.Spencer, M.Standish, E.Stephenson, A.Surtees, R.Swailes, B.Swann, J.Sweeney, G.Swire, L.Tattersall, A.Taylor, B.Taylor, J.Taylor, N.Taylor, S.Taylor, I.Thompson, J.Thompson, E.Thornley, S.Tibbs, C.Tilsley, D.J.Tinston, P.Titmuss, P.Tully, C.Turner, E.Turner, C.Unwin, J.Walsh, D.Wareham, B.Warren, N.P.Webb, R.Weir, B.White, J.B.White, K.White, S.White, D.Whiteley (Sheffield Museum), M.Willis, L.Wilson, M.Winstanley, J.Wood, K.Wood, T.Woodhouse, D.J.Woodward, G.Workman, P.Worthington, E.Wright, Yew Tree Junior School (Oldham), D.Zitman.

I am grateful to the Institute of Terrestrial Ecology, Monks Wood, Huntingdon for providing the historic records.

The distribution maps were produced using DMAP software. For more information about the software, contact:
Dr. Alan Morton, Blackthorn Cottage, Chawridge Lane, Winkfield, Windsor, SL4 4QR.

Finally, may I express my thanks to the Environment Agency and the Entomological Club for grants towards the publication costs.

PETER B.HARDY
Sale, Cheshire, December 1997.

INTRODUCTION

1. SURVEY METHODS

The 2 km square "tetrad" scale has been adopted in this work for the general coverage of the whole area. Scales such as 2 km or 10 km give a good idea of a species's overall distribution within a wide area such as a county which is too large for an intensive analysis, but they have their limitations (Hardy, 1994). For the smaller zone selected for more detailed mapping and analysis of butterflies' ability to utilise habitats within the contrast of urban and rural areas presented by Greater Manchester, the 7 X 5 km zone centred on the mid-Mersey Valley, the butterflies have been mapped at a smaller scale (1 km square) and the presence or absence of hostplant-habitats has been plotted. Even mapping to this scale has some disadvantages as it does not define accurately what percentage of a given area can provide suitable habitat for a species - a number of adjacent squares bearing a presence symbol can give the entirely false impression that the species occurs continuously all though the area. On the other hand, if using a smaller scale, say 100 m, there is the problem that unless one has a great deal of time to monitor the whole of the area very regularly and intensively, many of the squares will show blanks, and it will not be obvious whether this is due to there being no butterflies in them, or to their not having been visited sufficiently to produce an accurate record. Accordingly, a 3 X 2 km zone within the 7 X 5 block has been mapped at 100 m scale. Mapping at this level is useful for monitoring of resources and breeding - quantification - for fixed quadrats, to show the relationship between the butterflies and their hostplant-habitats in greater detail, and to allow comparison to be made between date periods.

(i) **Greater Manchester**. The mapping of the whole area on a tetrad scale (2 km X 2 km) includes records going back to 1980, though the majority are from the 1990s, when butterfly recording became more popular. Records have been supplied by a number of individuals and by such bodies as the Lyme Natural History Society, Wigan Field Club, Oldham and Bolton Museums and the Lancashire branch of the British Butterfly Conservation Society; these between them have provided extensive coverage particularly in the north-west and south-east of the mapped area. See the coverage maps, figs.7 and 8, which plot the number of known recording visits made to each of the 385 tetrads (fig.7), or 1411 1 km squares (fig.8) which fall wholly or partly within Greater Manchester. Clearly there is a limit to what extent of area can be covered in a given time by a given number of recorders, and as in any similar survey the results to some extent may reflect the distribution and extent of activity of recorders rather than the true range of the species, as shown by the map of all records of all species (fig.9). I have however made efforts to even this out. My own contribution to the survey is mainly the result of observations in 1993, 1994 and (especially) 1995. During those years I adopted a policy of going out recording on as many weekends as possible in the butterfly season, usually travelling by public transport to random or selected destinations, then walking for several hours and noting all the butterflies seen which could be identified. This was reasonably satisfactory as few of the species in this area present identification problems, apart from the perennial difficulty in distinguishing between the Small White *Pieris rapae* and the Green-Veined White *P. napi* in flight, mentioned in the accounts for those species. During those three years 1993-1995 I aimed to make as general as possible a coverage of the different habitats and to avoid concentrating on nature reserves and well-known sites. In 1996 and 1997 I aimed to fill some of the spaces not well covered by myself or others in the north and east, but even so I fear that this part of the area may have been slightly under-recorded compared with the remainder. Whilst certainly some of my walks were selected so as to include some potentially good habitat, I have nevertheless tried to record as fairly as possible in the less favourable areas which exist between the patches of prime habitat, and frequently have made walks totally at random. Above all, I have sought to treat the area as an interconnected whole; butterflies know no boundaries and certainly do not confine themselves to nature reserves.

I found the Geographers' A to Z Map Company's street atlas of Greater Manchester immensely useful for determining the grid references.

All my records from previous years have been included, going back to 1982; however for the majority of that time I was just making casual notes of butterflies which I happened to see and my records for the earlier years did tend to be biased towards easily accessible local sites or sites selected as "good" for wildlife.

Recording has been greatly facilitated by the fact that local authorities and warden services do much to promote exploration on foot. Most of the Manchester area is covered by a superb network of public footpaths and the vast majority of the sites are easily accessible. In many places, the river valleys in particular, and along the canals including where they pass through town centres and sites of heavy industry, it is possible to walk for a whole day hardly setting foot on a road. There will clearly be some gaps in the recording, however, as although so much of the area has excellent public access there are some potentially good sites which do not. The large grass expanse of the airport is a prime example of one of the areas impossible to record, and there are some other large inaccessible areas around factories. Many big gardens in housing areas, especially in districts like Hale or Bramhall, present a considerable amount of space which it is impracticable to survey completely and the best that can be taken is a sample, when recorders live in the area, or from such of the gardens as are visible from the roads. A few large tracts of open countryside in private ownership are inaccessible or of very limited access.

A map (fig.6) shows the approximate percentage of urban cover in each tetrad falling wholly or partly within Greater Manchester; this has been gauged from 1:50,000 Ordnance Survey maps.

(ii) **The 1 km scale maps of the mid-Mersey Valley**.

This zone comprises the whole of the former borough of Sale, Cheshire (prior to the 1974 local government reorganisation), plus adjoining parts of Urmston, Stretford, Chorlton, Wythenshawe and Timperley. For this zone, the butterflies have been mapped in greater detail, 1 km square level, and additionally an attempt has been made to assess the amounts of potential butterfly habitat within each 1 km square.

For the butterflies, data have been included, from my own records only, from the years 1992 to 1997, the majority being from 1994, 1995, 1996 and 1997, when I recorded in detail at a 6-figure level throughout each season; the records from 1992 and 1993 are mainly of the rarer species and also from a fixed transect which I monitored in 1993. It was not possible during the five years of observation to give enough coverage of each 100 m square during each species's flight period for maps at 100 m level to give a meaningful result; accordingly the maps have been drawn to show the *percentage* of the 100 m squares visited during the "butterfly season" (April, or March if there were any sightings that month, to October) in each 1 km square, in which each butterfly species was seen. For example, if in a particular 1 km square 25 of the 100 m squares were visited in the season(s) out of the possible 100, and a particular butterfly species was seen in 10 of them, this is treated as 40% occurrence.

For the hostplant-habitats, the zone was systematically surveyed during the winter of 1995/6. The aim was to note the presence or apparent absence of hostplant-habitat for each butterfly species in each square. Many sites were revisited during the 1996 and 1997 summers to check for those plants which might have been overlooked during winter when they were less apparent. The maps of the hostplant-habitats are presented at 100 m square level, as it is felt that sufficient coverage was made for them to be meaningful at this scale. It was not however found practicable at this initial stage to differentiate between the precise requirements of every butterfly species. Thus, the same map has been used for the two hesperiids and the three non-woodland satyrines as all these species feed on grasses. Even though there are definite differences between their requirements they overlap and currently insufficient data exist to discount the presence of one or more of them from any particular grassland biotope. Similarly, one map does duty for the four nettle-feeding nymphalines, another for the Large and Small White *Pieris brassicae* and *P. rapae*, and another for the Common Blue *Polyommatus icarus* and the occasional immigrant Clouded Yellow *Colias croceus*, which both feed on legumes. For the Speckled Wood *Pararge aegeria*, which has very different requirements from the other three satyrines which occur in this region, the map shows areas of mature and semi-mature trees assessed as likely to provide the shade requirements for the butterfly; the presence of enough grass for breeding has been assumed as likely in any such area.

Maps are provided showing the approximate percentage of urban cover (i.e. built-up areas

- houses and gardens, offices, shops and industrial areas) as against open "green" space in each 1 km square (figs.77 and 78), along with a detailed assessment of areas of rough grass and associated semi-natural ("wild") vegetation which appear to provide potential habitat for the majority of butterfly species (fig.74), and also the extent of the *unsuitable* green space - grassland which is either too severely grazed to support butterfly larvae, or constantly mown, such as parks and playing fields (figs.75 and 76). Finally, coverage maps are presented showing the squares surveyed at any time during the whole year (fig.79) and during the "butterfly season" (fig.80). For the purposes of these maps, a visit to any individual square means a pass through any part of that square on a day when potentially butterflies could have been seen, and does not necessarily indicate a thorough coverage of the square. It must also be borne in mind that it is not practicable to cover all squares with equal thoroughness; a square falling wholly within public open space can be covered completely whereas when a square is in a built-up area one is confined to the public roadways.

It is appreciated that the hostplant/habitat and butterfly species distribution will not be static. Indeed, a widening of the M-63 motorway, across the whole width of the 7 km X 5 km zone, which is expected to start within the next two years, will cause very great changes. One reason for the selection of this zone for more detailed mapping was to be able to compare the impact of such environmental changes as motorway development.

(iii) **The mid-Mersey Valley: 100 m scale maps.**

For this block (grid squares SJ7892, SJ7893, SJ7992, SJ7993, SJ8092 and SJ8093) the butterflies have been mapped to 100 m square level and the assessment of hostplant/habitat has been carried out in greater detail, and also an attempt has been made to map the main nectar sources (figs.163 to 168).

Assessments have been made of the percentage of each 100 m square occupied by what appeared to be potential hostplant-habitat (figs.118 and 119). Also, the percentage occupied by unsuitable or marginally suitable habitat has been assessed - open spaces such as regularly mown or heavily grazed grass (for example, playing fields or horse-riding areas respectively) which would appear on a map as "green" space; and a more detailed analysis of built-up areas has been attempted, with separate figures for houses/gardens, offices/shops/industry, and areas largely covered by roads, as well as areas of crop-growing, i.e. farms and allotments, and water-side areas (see figs.120-125). In this 3 x 2 km zone every 100 m square, apart from five (SJ799929, SJ797931, SJ798930, SJ799930 and SJ800929) wholly under the lake in the Sale water-park and three (SJ781932, SJ781933 and SJ782932) wholly within the Stretford sewage farm, which it was not possible to obtain permission to enter, has been visited during the "butterfly season"; however it must again be remembered that some squares could be covered more thoroughly than others; some of the squares in open countryside have full public access and could be covered very thoroughly, whereas others mainly on farmland could be given at best a very superficial coverage from the field edges; and in squares in suburban areas coverage was normally limited to what was visible from the public roadways and parks, and for obvious reasons when recording in a street of houses, especially if it is one which comes to a dead end, one tends not to loiter in any one spot staring into the front gardens for too long!

The butterflies are recorded as single or multiple sightings, within each 100 m square. Again, for this 3 X 2 km zone I have used my own records only; most are from 1994, 1995, 1996 and 1997, plus some from 1993 when I mapped certain species on a fixed transect (Hardy, 1994) and a few 1992 records of the scarcer species (*Colias croceus, Gonepteryx rhamni, Celastrina argiolus* and *Pyronia tithonus*); however most of my records prior to 1994 were at 1 km level and therefore could not be included.

I made most of the recordings in this zone on Mondays to Fridays during office lunch breaks ("flexi-time" permitting these to be extended up to 2 hours being a definite advantage!), usually reserving the weekends for recording further afield. Although in earlier years I admit to often having preferentially selected the most likely routes for my walks, in 1996 and 1997 I made an effort to even this out and ensure that the less obvious parts of the zone were adequately covered. Even so, it is inevitable that the squares closest to my house and office, and also those lying between my house and the local shops, received more coverage than the peripheral squares.

(iv) **1996/7 nectar survey**.
During the 1996 and 1997 seasons, a record was kept of the number of days on which each butterfly species was seen using any particular plant species as a nectar source (table 2, page 48). Most observations were in the Mersey Valley but some were from other districts, especially sites within a mile of Manchester city centre where I made a number of observations in 1996. Additional records for 1996 were provided by K.Butterworth from the Oldham area. Analyses by plant and butterfly species are tabulated (table 2, page 50). Distribution of the six most frequently used nectar plants has been plotted for the 3 X 2 km zone (figs.163 to 168).

2. MANCHESTER ENVIRONMENTS.

It is easy to think of "Greater Manchester" as an almost wholly built-up area and practically useless for butterflies. This is not the case. As well as housing, offices, shops and industry, within the administrative entity are a broad range of habitats: mature woodland, scrubland, grassland, high moorland, mossland, agricultural land, lakes, wetlands, river valleys, parkland, canal sides, railway and motorway embankments and cuttings, suburban and city gardens and green space, waste land within and around industrial areas, and innumerable combinations. In the south-west starts the fertile, low-lying agricultural expanse of the "Cheshire Plain"; extensive rolling agricultural land occurs around Wigan also, continuing northwards into mid-Lancashire. Also in the west are the low-lying "mosses", albeit for the most part sadly altered by drainage and peat extraction. In the north, and notably to the east, are the Pennines and south Lancashire hills; these include high open moorland, for the most part very barren and degraded by sheep-grazing but nevertheless providing some pockets of habitat, chiefly east of Littleborough, Oldham, Greenfield and Stalybridge. At lower altitude, much of the area formerly in Lancashire was once dominated by industry, especially cotton milling; some of this remains but the one-time image of "dark satanic mills" is now largely a thing of the past; not only have many former industrial areas now been completely converted to parkland but also green areas abound alongside the remnants of this industry. Greater Manchester Council, prior to its disbandment, promoted the cleaning-up of river valleys and creation of country parks and walkways; although the parent body no longer exists the good work continues under the various warden schemes, notably in the Mersey, Croal/Irwell, Medlock, Tame, Etherow/Goyt and Bollin Valleys.

On the other hand, newer industrial areas (or "business parks"/"science parks" as it seems to have become customary to call them) continue to be built in some areas (for example, Salford Quays (SJ8097)), and whilst they are not as "dark" or "satanic" as former industry such green space around them as exists contains little natural vegetation.

It is difficult to assess what the natural vegetation of the area really is as so many alien plants have been introduced or have become naturalised. A very large proportion of the land within the conurbation is occupied by suburban and urban gardens, and in spite of the recent publicity given to the possibilities of "gardening for wildlife" and wild-flower gardening, the vast majority of the flowering plants, trees and shrubs grown in these gardens are non-indigenous. The same is still true of most formal parkland and public gardens; and although some local authority-controlled warden services have planted native trees and shrubs in and around the river valleys, sadly the majority of "amenity" landscaping around newly "re-developed" sites is still with exotic trees and shrubs. Abandoned sites, left untouched for a few years, are probably the nearest biotopes available to a natural environment; these regularly develop a mixture of grass and wild flowers including clover, sorrel, nettle and thistle, which between them support fifteen of the butterfly species recorded in the area; crucifers which are pioneering plants on disturbed land support a further four species, and other flowers colonise which serve as nectar sources.

Some butterflies have however benefited by the introduced flora: alder-buckthorn which supports the Brimstone *Gonepteryx rhamni* and the holly and ivy used by the Holly Blue *Celastrina argiolus* have mostly been artificially planted. The occurrence of these two butterfly species in Greater Manchester is largely a result of man's introduction of their hosts. The Common Blue *Polyommatus icarus* would also probably be absent or very scarce if outside elements, particularly limestone, had not been artificially added to the geology.

As well as indigenous vegetation, a number of "garden escapes" and other naturalised plants grow wild, especially on regenerated sites in and around the city. Invasive "aliens" such as Himalayan balsam *Impatiens glandulifera* and Japanese knotweed *Fallopia japonica* are all too familiar as increasingly dominant elements of our flora, notably along the river valleys. Less aversion tends to be felt towards the wild buddleias *Buddleja davidii* which rapidly colonise demolition sites, railway sidings and other waste land in the urban areas, or the vast swathes of Michaelmas daisies *Aster novi-belgii* which have spread, especially, over former tips - probably because both these plants remain popular as garden flowers. The table of nectar sources recorded in 1996/7 shows these two plants as among the six most frequently used by butterflies, with buddleia appearing a clear first. It should however be borne in mind that it is primarily the five "aristocratic" nymphalines (the Red Admiral *Vanessa atalanta*, the Painted Lady *V. cardui*, the Small Tortoiseshell *Aglais urticae*, the Peacock *Inachis io* and the Comma *Polygonia c-album*) and the most mobile of the pierids (the Large White *Pieris brassicae* and the Small White *P. rapae*) which regularly use it; other species were only recorded on seven occasions during the two years. It is possible to over-estimate the usefulness of the plant as a nectar source by omitting to take account of three factors: (a) the vagility of the nymphalines and pierids, (b) the extent to which they move away from breeding habitats in search of nectar, including into gardens which contain no breeding habitat (compare the 3 X 2 km maps for *V. atalanta* (fig.139), its hostplant nettle (fig.160), and its most favoured nectar source buddleia (fig.164)), and (c) the general apparency and popular appeal of these "showy" butterflies to the general public.

Much popular literature has been published regarding the apparent value of certain "garden" flowers as nectar plants for butterflies (Newman (1967), Rothschild & Farrell (1983), Vickery (1990, 1991a, 1991b, 1992, 1993, 1994, 1995, 1996). The present survey has shown that besides the above two non-indigenous plants (data for wild and cultivated examples combined) the most valuable nectar sources are three plants generally regarded as insidious weeds - creeping thistle *Cirsium arvense*, ragwort *Senecio jacobaea* and dandelion *Taraxacum officinale* agg. - and another popular source, bramble *Rubus fruticosus* agg., is a plant which, although native and valued by some of the general public on account of its tasty fruit, still tends to be thought of as an invasive nuisance by reserve wardens. During 1996, the only other non-native plants with an appreciable number of records of nectaring butterflies were Oxford ragwort *Senecio squalidus* and garden privet *Ligustrum ovalifolium* (when it is allowed to flower). During 1997 two other introduced plants, both "garden" ones, drew a number of records, these being Lavender *Lavandula officinalis* with 16 records and Ice-plant *Sedum spectabile* with 17; the last-mentioned plant was especially attractive to the unusually abundant *A.urticae* that year. Apart from French marigold (7 records) no other "cultivated" or "garden" flower drew more than a single record in 1996, or more than two records in 1996/7.

3. SPECIES RICHNESS AND DISTRIBUTION

Twenty-seven butterfly species have been recorded during this survey. I have recorded twenty-three myself (twenty of these have been seen in the Mersey Valley) and a further four (the Purple Hairstreak *Quercusia quercus*, the Long-tailed Blue *Lampides boeticus*, the Camberwell Beauty *Nymphalis antiopa* and the Dark Green Fritillary *Argynnis aglaja*) have been reliably reported by other observers. Of these, *A. aglaja* can at best be considered an occasional stray, the occurrences of *N. antiopa* in 1995 were part of an exceptional nationwide influx of this scarce migrant that year and are unlikely to be repeated, and *L.boeticus* was recorded from a single sighting in 1995, interestingly at the same time as the *N.antiopa* records. The Clouded Yellow *Colias croceus* is a very irregular migrant and has only been seen in a few years during the survey period, most of the sightings having been in 1983 and 1992. Of the resident species, the three Hairstreaks (the Green *Callophrys rubi*, the White-Letter *Satyrium w-album* and the Purple *Q. quercus*) are very local and scarce; the Holly Blue *Celastrina argiolus* and the Brimstone *Gonepteryx rhamni* have normally been scarce and unpredictable, though there are signs of a recent increase in range and abundance. Two of the satyrines are of restricted distribution, the Gatekeeper *Pyronia tithonus* mainly in the west (but steadily increasing its range) and the Small Heath *Coenonympha pamphilus* in the higher ground to the east. The remaining sixteen species (including the two common migrants, the Red Admiral *Vanessa atalanta* and the Painted Lady *V. cardui*) are widely distributed.

The coincidence map (fig.9) shows the number of species recorded in each tetrad. The maximum number of species recorded in any tetrad is 21; in 11 tetrads 20 or 21 species have been recorded: SD5202, SD5606, SD5802, SD5806, SJ6298, SJ6698, SJ7292, SJ7892, SJ8082, SJ8484 and SJ8886. These tetrads are all in the south and west of the area and were in the districts with highest recorder coverage (mainly by the Wigan Field Club, K.McCabe, B.T.Shaw, S.H.Hind and myself); see figs.7 and 8. Tetrads containing a reasonable percentage of rough ground and woodland can normally be expected to harbour at least sixteen species. Even SJ8498, the tetrad including Manchester city centre, has fifteen species.

During the survey, it became evident that "wild" and regenerating areas, with little or no "management", are by far the richest habitats. It also became apparent that very often islands of habitat amidst industrial areas are *richer* than much of the open countryside. The butterflies' powers of searching out and locating small "islands" of habitat amidst heavily built-up areas should not be underestimated (compare Dennis and Shreeve (1996) and Owen, (1949)). The natural flora which develops in these inner-city sites includes "weeds" which serve as larval hostplants, and others, notably ragwort and thistles, which provide adult nectar sources. As mentioned in the "Environments" section, in many of these sites non-indigenous "garden escape" plants, especially buddleia and Michaelmas daisy, have assumed an even greater importance as nectar sources. Agricultural "improved" fields, which comprise a large proportion of the open countryside, are not good habitats. Neither, in spite of literature suggesting otherwise, are gardens. Some contributors to the survey clearly record with a bias on their own gardens. Year-round records for a single garden may well produce quite an impressive list with more records, of species and individual butterflies, than a single day's recording in a good semi-natural habitat (an example being a single garden in Bramhall (SJ9086) in which a total of 19 species have been recorded during the survey period) and thus give the totally false impression that the former is the richer habitat. I grant that butterflies do occasionally breed in gardens especially if a few "weeds" are allowed to grow, even very small gardens (including my own - see the species account for *Lycaena phlaeas*, and R.L.H.Dennis of Wilmslow, Cheshire, just outside the mapped area, reports that five species breed in his garden - *P.rapae, P. napi, A.cardamines, P.aegeria* and *C.argiolus*). I accept that when making a recording walk through any given area and aiming to cover a cross-section of habitats, the recording in open countryside is easier than in the suburbs and that a large proportion of a suburban area may comprise back gardens in which there may be overlooked habitat; the larger the gardens, the greater the proportion of the area which it is impossible to survey, but even so I feel that the vast majority of the butterflies seen in gardens have been drawn into them seeking nectar sources or are on passage, and have bred elsewhere, and my observations in this survey bear this out. Year-round records from an individual garden give the false impression that those butterfly species which frequently come into gardens, especially the large nymphalids, are much commoner than they actually are in relation to other species which tend to remain in the breeding habitat. Cemeteries and formal parks, although they form large "green lungs" in built-up areas, are also very poor habitats unless they contain areas where the grass and wild flowers are allowed to grow. Recently, patches in some inner-city parks, including Alexandra Park (SJ8395) and Birchfields Park (SJ8594), have been allowed to regenerate naturally, with spectacular results. Some of the allotments which are popular in suburban areas provide better habitats than gardens; the *Brassica*-feeding pierids are the species which benefit most but other species also find a home here. Even the most unlikely places are worth trying, as sooner or later some butterflies will visit them.

4. DISTRIBUTION CHANGES.

i. **Historic records.** The maps at 10 km scale (figs.37 to 72) are from historic records obtained courtesy of the Institute of Terrestrial Ecology, Monks Wood, and with the addition of a record of *Callophrys rubi* from Botany Bay Wood in the early 20th century, supplied by Bolton Museum. It has only been possible to plot to this scale as recording prior to 1980 was nothing like as intense as today and most of the records only show two-figure grid references. The lack of any records in grid square SD91 may safely be taken as due to lack of recorders/collectors in the Oldham area, not absence of any butterflies. In some cases where the locality is merely given as "Manchester", the Institute of Terrestrial Ecology have presumed it to be in SJ89, the 10 km square including the city of Manchester; these records should be interpreted with caution as in old collections specimens were frequently labelled with the name of the nearest town and the exact locality could have been some distance away. The possiblility of misidentification of rarer species (*C.minimus* and *A.artaxerxes*?) should also be borne in mind.

The records are plotted as follows: the symbol # indicates that the species has only been recorded from the square prior to 1900; open circles indicate 20th-century records prior to 1940 only; closed circles indicate records up to 1979. In some of the squares which are partly inside and partly outside the present-day Greater Manchester boundary, the site is not specified in the record and it is therefore uncertain whether or not it was within the area; in these instances a half-size symbol is shown. Whilst many species appear to show a distribution similar to the present day, it is difficult to gauge an accurate comparison at this scale of mapping which at best can only give a very general idea of a distribution pattern.

The picture which seems to emerge is however as follows: During the periods prior to 1980, and in particular from 1940 to 1979, of the 27 species recorded during the present survey, 10 species were generally distributed and reasonably common throughout the area: *P.brassicae, P.rapae, P.napi, L.phlaeas, V.atalanta, V.cardui, A.urticae, I.io, L.megera* and *M.jurtina*. 9 species were noticeably less widely distributed than at present: *T.sylvestris* (virtually unknown), *O.venata, G.rhamni, A.cardamines, P.icarus, C.argiolus, P.c-album, P.aegeria* and *P.tithonus*. 2 species which appear to have been much more widely distributed prior to 1980 than at present are *S.w-album* and *C.pamphilus*. *C.rubi* was recorded in some southern squares in which it appears to be absent today, yet there were no records of it from the north and east; this is however more likely to reflect the absence of recorders in the north and east than a range-extension. *Q.quercus* appears to have been very rare or overlooked; and 3 species, all "strays", were unknown in the area prior to 1980: *L.boeticus, N.antiopa* and *A.aglaja*.

The following are species which have been recorded historically from the area but for which there is no record from 1980 onwards.

Dingy Skipper *Erynnis tages* (fig.39), old records from SJ79 Chat Moss, 1905), SJ88 (Cottrell Wood, 1905) and SJ89 ("Manchester" 1857+), and a more recent record from SJ98 (site unspecified, 1978).

Wood White *Leptidea sinapis* (fig.40), pre-1940 records from SJ78 (site unspecified, 1939+) and SJ98 ("Parkgate", 1903 - this record would appear extremely doubtful as there is no locality with this name on present-day maps of this square, the nearest such is on the Wirral).

Small Blue *Cupido minimus* (fig.51), a record from SJ89 ("Manchester area", 1908).

Silver-Studded Blue *Plebejus argus* (fig.52), records from SJ79 (Chat Moss, 1908) and SJ89 ("Manchester", 1857+).

Northern Brown Argus *Aricia artaxerxes* (fig.53), records from SJ79 (Chat Moss, 1910) and SJ89 ("near Manchester", 1910+).

Large Tortoiseshell *Nymphalis polychloros* (fig.59), a record from SJ89 ("Manchester", 1857+). A record from SJ78 in 1899 is outside the area.

Small Pearl-Bordered Fritillary *Boloria selene* (fig.62), a record from SJ89 ("Manchester", 1857+).

Pearl-Bordered Fritillary *Boloria euphrosyne* (fig.63), a record from SJ89 ("Manchester", 1857+).

Silver-washed Fritillary *Argynnis paphia* (fig.64), a single record from SJ78 in 1950; the site is unspecified and it is uncertain whether or not it was inside the present Greater Manchester boundary.

Marsh Fritillary *Euphydryas aurinia* (fig.65), records from SJ78 (site unspecified, 1939+), SJ88 (Stockport, 1846), SJ89 ("Manchester", 1857+) and SJ98 ("near Stockport", 1845).

Grayling *Hipparchia semele* (fig.68), a record from SJ98 (site unspecified, 1939+).

Ringlet *Aphantopus hyperantus*, no confirmed records from within what is now Greater Manchester. A record from SJ78 in 1976 was outside the area (Rostherne Mere). In the early 1990s there was a rumour that this species might be breeding in the vicinity of Partington (SJ7291); it could not be verified.

Large Heath *Coenonympha tullia* (fig.72), records from SJ79 (Chat Moss, 1857+) and SJ88 (site unspecified, 1939+); this species was originally known as the "Manchester Argus" but that name has long since ceased to be appropriate and it probably became extinct in the area prior to 1920 (Rutherford (1983) and Whitehead (1986).

ii. **Recent records.** In more recent times the picture is encouraging. There is no doubt that butterflies, and other wildlife, have benefited greatly from the deliberate cleaning-up of the river valleys and some canals; these waterways form a hub of flyways through which mobile wildlife can penetrate practically to the city centre. Greater Manchester now contains a number of nature reserves (see map, fig.4), most of which are in the river valleys, and although none, as far as I am aware, are specifically "managed" for butterflies, they are certainly beneficial. Greater Manchester Council's tree-planting scheme in the late 1970s created numerous woodland and woodland-ride habitats which have benefited many butterfly species and especially assisted the spread of the Speckled Wood *Pararge aegeria*. I would however like to make the point that as well as designated reserves there are many other habitats, less obvious but every bit as important. I refer to those which have not been intentionally created by man but which have *evolved*, through natural succession, largely as a result of abandonment of former environmentally destructive activities and subsequent benign neglect. So-called "waste" land is far richer in invertebrate life than formal parkland. Far too often these scraps of "waste" land are regarded as eyesores which need building over or at least "tidying-up"; fortunately, although many such sites are short-lived, new ones keep appearing as a result of demolition, abandonment of former industry and the like. Many of the railway lines abandoned in the 1960s have turned into linear strips of first-class habitat; more recently other habitats have developed, or are in the process of developing, from former railway goods yards, or abandoned gas works and power stations, often quite close to the city centre. With changes in fuel, and with waste-disposal having become so much more efficient, many former rubbish tips, and in the west of the area coal mines, have gone "back to nature", and in many cases have introduced a new geological element and associated flora and fauna quite different from the surrounding countryside. This is especially the case where limestone has been introduced, as has occurred by way of railway ballast, tips containing roadstone, hard standings around the foundations of demolished buildings, and the like. The Common Blue *Polyommatus icarus* is a species which has benefited as a result. In some places habitats have developed from abandoned sports fields which have revegetated naturally for a few years before being reclaimed for building, and from former tree nurseries. Even very small habitats, tiny scraps of revegetating waste, are worth searching. Again, small semi-natural sites surviving in an otherwise hostile environment should not be overlooked; in a small relict wood surviving amidst formal playing fields in Brooklands Estate (SJ7989) I recorded nine species of butterfly within just over an hour in June 1995.

Comparing the present-day butterfly fauna with that which I remember from childhood (in the late 1950s), especially the Sale area (covered by the 7 X 5 km maps), the results are surprising. The current species total in that area is 20 - *T. sylvestris, O.venata, C. croceus, G. rhamni, P. brassicae, P. rapae, P. napi, A. cardamines, L. phlaeas, P. icarus, C. argiolus, V. atalanta, V. cardui, A. urticae, I. io, P. c-album, P. aegeria, L. megera, P. tithonus* and

M. jurtina. I was familiar with every open space in Sale in 1959 but can only recall a maximum of 8 species. The three *Pieris* species were common, *A. urticae* reasonably so, and the migrant nymphalids *V. atalanta* and *V. cardui* occurred unpredictably then as now. I knew sites for *L. phlaeas*, mostly in Stretford; *M. jurtina* occurred in the Altrincham district. The others apparently did not occur. For instance, each spring I would search likely places in vain for *A. cardamines*. Regrettably I did not keep detailed notes then and my memories *may* be clouded with time; however the close resemblance between them and the contemporary report from Blackie (1946) of ten years of observation from nearby Wythenshawe suggests that they were reasonably accurate. Blackie mentions the three *Pieris* species as common, a few *A. cardamines* seen in 1945 but none previously or in 1946; of the nymphalines he quotes *A. urticae* and *V. atalanta* as common, *V. cardui* as occurring in certain years and *I. io* as rare. The only lycaenid definitely present was *L. phlaeas*, although Blackie reported possible sightings of single *Q. quercus* and *C. argiolus*; and *P. icarus* had apparently become extinct following a single sighting in 1937. "A few" *M. jurtina* had occurred, though not every year - the species "did not seem to be securely established and its prospects did not look very good"; and he had found *L. megera* in 1943 and 1945 only. No Hesperiidae occurred at all. He also mentioned a single sighting of a Fritillary briefly nectaring on a garden variety of *Achillea millefolium* - "it was beyond doubt either *Argynnis cydippe* L. or *A. aglaia* L., but I had no opportunity of seeing the upperside and thus of putting the matter beyond doubt". Blackie's sighting was almost certainly a wandering *A. aglaja* from the Peak district, similar to the strays in 1992 and 1993 mentioned below in the account for that species.

Over the 40-year period since my childhood observations there has been a reduction in air pollution as a result of the clean air acts, and an improvement in the climate, which may well have been factors contributing to the species increase (Hardy & Dennis, 1997). It is also probable that in this region there has been a net gain in suitable habitats. True, some open space has been lost to building (compare the 7 X 5 km urban cover maps for pre-1960 and 1996, figs.77 and 78), but there have been changes in the *suitability* of the open space. In the 1950s the concept of deliberately designating or "managing" land specifically for wildlife was unknown, and vast areas of the Mersey Valley were under rubbish tips or low-quality agricultural land. With more efficient methods of waste disposal nearly all the tips have gone, and so has the former Withington sewage works (SJ8093, actually in Chorlton). Habitats in this sector include a number of nature reserves in the control of the Mersey Valley Warden Service, notably the extensive woodland of the Chorlton Ees (developed from the former sewage works), smaller woodlands near the Priory, Sale, and the Mersey Valley Visitor Centre (both SJ8092), and the artificially-created wetland of Broad Ees Dole (SJ7993, 8093), adjacent to the Sale water-park. These sites are "managed" mainly for their birds and botanical interest, but all wildlife, including butterflies, benefits incidentally. Other "managed" habitats include hay meadows, horse-riding areas, and a scrubland set aside for "earth magic" activities. Several sites have developed from former tips: the extensive Stretford tip (SJ7893/4) is now a rolling, open grassland; others, notably the Bethell's tip (SJ7992) adjacent to the Priory have re-vegetated with scrub and species-rich grassland with limestone flora.

The situation in the Mersey Valley is likely to be typical of much of the Manchester area. Although detailed records from 40 years ago are unavailable, observers from many parts of the conurbation have noted similar increases in the same species as in the Valley, including the phenomenal spread of *T. sylvestris, P. tithonus* and *P. aegeria*, in the 1990s (Hardy, Hind & Dennis (1993) and Hardy & Dennis (1997)). Whatever the factors causing these range extensions are, they affect the whole area and not just the Mersey Valley.

5. CONSERVATION

The butterflies of the Manchester area are not national rarities. They are not what is usually understood by the expression "endangered species". Given a choice of anywhere in the country, we would be very unlikely to select a site in the Manchester area as an ideal place to introduce a visitor from abroad to Britain's butterflies.

But does the fact that Manchester's butterflies are mostly common species make them any less worthy of conservation? I think not. It is of very great interest that 43% of the British butterfly species can, and do, successfully find somewhere to survive in this "metropolitan" area. Not everyone has the means to travel far to look for butterflies, and the butterflies of species-poor areas near home are every bit as worthy of study as those more numerous further afield. I believe that every child growing up in and around Manchester should have the opportunity to see and appreciate the butterflies close to his/her home. The message then needs to be got across that the lovely butterflies on the garden buddleia would not be there if they did not have somewhere to *breed*.

It is, sadly, a fact that most of Manchester's butterflies are as widespread as they are as a result of their at times extraordinary powers of adaptability and not because we have deliberately helped them. Species-rich habitats evolve slowly, but can be destroyed very quickly, and this still all too often happens. An example was the first-class hay-meadow at Bruntwood, Cheadle (SJ8587) which disappeared under a Sainsbury's supermarket in spite of valiant efforts by the Manchester Wildlife Trust to save it. Another example is the vast expanse of regenerated former farmland at Dumplington (SJ7696, 7796) which at the time of writing (1997) is currently being covered by a "shopping city". It is heart-rending to see a site which had developed into a prime wildlife habitat wiped out by bulldozers in a matter of weeks.

Sites rarely remain the same for very long. The following are some instances of habitat changes in and around the Mersey Valley. An interesting area of young woodland with associated edge and ride habitats, combined with the adjacent river-bank and motorway embankment, developed on the north side of the Crossford sports field (SJ7993); as regards Lepidoptera this habitat had its heyday in the mid-1980s, but the ride and embankment have since become rather shaded as vegetation succession was not checked by "management"; fewer butterflies now use it, and the Narrow-Bordered Five-Spot Burnet moth *Zygaena lonicerae*, which formerly abounded there, has died out. (The increased shade has however benefited one butterfly species, the Speckled Wood *Pararge aegeria*, which has colonised the site since 1994). Also in this sports field is a noteworthy habitat formed by a north-south drain between the west and east sections of the field. The latter section is on a slightly higher level so that the edge of the drain forms a west-facing slope and has a warm microclimate attractive to overwintered nymphalines in the early days of spring; nettles growing along the drain provide a breeding site (Hardy, 1995). Regrettably, another sports field nearby on Glebelands Road (SJ7892), which was abandoned and during the late 1980s had developed into a fine habitat for several species, was lost to building.

An excellent site, just outside the Valley proper but within the mapped area, was a long strip of rough grassland, running south-west to north-east, on the Sale-Wythenshawe boundary (SJ8090), which included an abandoned tree nursery and in quite a small area contained a full range of habitats from open grassland to near-mature woodland, and their associated butterfly species. This site may be cited as an illustration of how, to assess accurately the value of a site to wildlife, it is necessary to make a number of visits at different times of year: on 6.v.1995 only one butterfly species (*P. rapae*) was recorded there, on 5.vii.1995 six species and on 13.viii.1995 six species - not all the same - were seen. Eight species were recorded on 6.viii.1994, making the total of species recorded at the site to twelve (*T. sylvestris, O. venata, P. brassicae, P. rapae, P. napi, L. phlaeas, V. atalanta, A. urticae, I. io, P. aegeria, L. megera* and *M. jurtina*). Unfortunately as with so many others, this site was destroyed for building in 1996; as none of its resident butterflies were rarities no case could be made out for its retention on these grounds despite its exceptional interest.

Perhaps the most noteworthy of all the habitats which have fortuitously evolved in the Mersey Valley is the river-bank itself, which has been artificially built-up in the form of a "levee" to prevent flooding. For a number of years I have regularly monitored the stretch of the north bank to the west of the A.56 road on the Sale-Stretford boundary (SJ7892, 7893, 7993), where it forms a linear south- or south-west-facing slope. As well as providing the warm microclimate favoured by most of the resident butterflies the bank acts as an ideal "flyway" for patrolling and migrant species and over the years of the survey eighteen of the total of twenty species recorded in the mid-Mersey Valley have been observed along this stretch. In some years the butterflies along this bank have appeared in extraordinary profusion. An entry in my diary for 4.viii.1984 reminds me that on that day, feeling rather depressed at just having heard that I had been unsuccessful in a bid for promotion at work, I was "cheered immensely" by "myriads" of butterflies all along this bank, of eight species (*P.brassicae, P.rapae, P.napi, L.phlaeas, P.icarus, A.urticae, L.megera* and *M.jurtina*); the Common Blue

P.icarus was extraordinarily abundant on the bank at that time. Although in 1986-7 the vegetation on this section of the bank was almost completely destroyed as the water authority found it necessary to lower the berm and re-shape and strengthen the levee, the bank settled down fairly quickly after completion of the work and within three years its species-richness had almost reverted to its former level. In 1992, apparently as a result of a combination of favourable climatic factors and management resulting in the optimum balance of hostplants and nectar sources, another spectacular wave of abundance occurred, and the Clouded Yellow *C.croceus*, making one of its rare migratory visits to this part of the country, found ideal breeding conditions. Again referring to my diary, I find that on 16.vii. I was "roused considerably" by a lunchtime walk along the bank observing *C.croceus* along with the first ever record of the Gatekeeper *P.tithonus* from Stretford, along with "hordes" of *A.urticae* and numbers of *T.sylvestris, P.brassicae, P.rapae, P.napi* and *M.jurtina*. Other parts of the bank continue to undergo this periodic strengthening; and, additionally, the flora and fauna of the whole bank have to contend with an ongoing lesser disturbance by mowing, which the Environment Agency (water authority) carry out to make the bank look tidier to the public who use the public footpaths along it as well as strengthening the roots of the grass and thus stabilising the bank. The water authority formerly adopted a policy of indiscriminately mowing the vegetation at intervals ranging from one to three times per year; following discussion they have become a considerably more conservation-minded and now normally confine the mowing to the lower section below the berm, with a single autumn cut on the upper section.

In the early 1970s the Mersey Valley underwent considerable disturbance because of the building of the M.63 motorway through it. Since then, it gradually settled down into the mosaic of habitats described above. Now, sadly, it stands to be extensively disturbed again by a widening of the motorway and construction of further access roads. The whole width of the 7 X 5 km mapped area stands to be affected. The Highways Agency have drawn up plans for "restoration" of damaged wildlife habitats and have included proposals relating to the butterflies (Highways Agency, 1995); however these proposals were not the work of an expert lepidopterist. It remains to be seen just how catastrophic the effect on the butterfly populations will be and how soon they will recover; hopefully a future survey will cover this.

The pattern of events in the Mersey Valley is typical of many parts of Greater Manchester. The continuing trend towards private transport results in more and more schemes for building new roads and widening existing ones, and development of new industrial sites and shopping precincts, sometimes in what had appeared to be sacrosanct "green belt" locations.

Looking to the future, and especially with regard to Greater Manchester as a whole, it is unlikely that the net gain in areas of suitable habitat will continue. The environmentally-conscious "green" movement of the 1970s and 1980s seems sadly to have passed its peak, and in many minds "familiarity has bred contempt" for the often-repeated conservation message. Cutbacks in local authority expenditure on warden services and environmental education, along with the ever-increasing trend towards "market-testing" and privatisation, result in every available scrap of land being looked at with a calculation of how much profit it could generate if put to commercial use. There is also too much of a tendency to regard patches of naturally revegetated land as "eyesores" needing "tidying up", a reluctance to let Nature carry out the restoration and an obsession with artificially "restoring" disturbed ground by planting, often with totally inappropriate vegetation; land which escapes being built over risks being converted into manicured grass or planted with non-native trees and shrubs.

Conservation also takes second place to recreation. Far more of the general public regard areas of green space in and around residential districts as of more "use" as playing fields or golf courses, or even as plain expanses of closely-mown grass between blocks of flats, than as wildlife haunts. The future of Manchester's butterflies depends to a large extent on whether the ever-increasing human population can learn to accept that some patches of less man-modified vegetation are desirable.

Open land around Manchester, especially in the less affluent districts, is also subject to abuse such as rubbish-dumping, motor-cycle "scrambling" and shooting, and, sadly, the sites where these activities occur are often some of the best wildlife habitats. Some control over these practices by countryside wardens has certainly been achieved; it is also accepted that they are not as harmful to butterflies as to the more easily disturbed wildlife such as birds; nevertheless they are indicative of a mindless attitude which has no regard for the environment and apparently prefers to destroy other living things rather than conserve them. Yet another threat to habitats, a less obvious one, lies in making them too obvious. Whilst some improvement of public footpaths and the creation of signposted walks are undoubtedly good, it can happen that widening and resurfacing of paths to make them more attractive to persons who would otherwise be unable or reluctant to use them may significantly reduce the width of available habitat, and is frequently accompanied by severe moving of the path-side vegetation, and trampling and cycle wheels have a further impact, made worse still when the path becomes wet.

To many of the general public, a suggestion of taking the family somewhere in the Manchester area to see some butterflies would be answered with a recommendation to visit the Bolton butterfly house. Possibly a few who had seen leaflets advertising walks in one of the river valleys might suggest one of the nature reserves. It is very unlikely that any would suggest the nearest scrap of "waste" land - but this would probably be the cheapest and most convenient option, as well as the one giving the truest idea of the ecological requirements of butterflies. During my wanderings with a recording notebook I have frequently attracted the attention of onlookers curious to know what I was doing, and drawn their attention to the existence in their neighbourhood of numbers of butterflies of which they were hitherto unaware.

If this booklet can generate more appreciation of butterflies in an urban environment, their requirements, their adaptability, their value as indicator species for wildlife in general, and thus even a slight shifting of views as to the value of land in or reverting to a semi-natural state, it will not have been written in vain.

SPECIES ACCOUNTS

The notes in the following species accounts relate solely to observations made during the survey in Greater Manchester (or in a few instances just outside the boundary) and draw attention where appropriate to differences from the same factors in other areas.

Family HESPERIIDAE

SMALL SKIPPER, *Thymelicus sylvestris* (Poda, 1761)

Habitats. Warm sites with rough grass, abundant nectar flowers and, usually, some scrub. Many suitable sites are on road and railway verges, abandoned railway yards and other regenerating patches of waste ground including those close to the city centre and industrial areas. The species is less ecologically tolerant than the Large Skipper *Ochlodes venata*, requiring warmer, more sheltered habitats. For instance, at a site on the east side of the A.538 road, just north of the tunnel under the airport runway (SJ8083) a steep west-facing slope, created by excavating the cutting for the road, has developed into ideal habitat for the species: during a late afternoon visit on 10.vii.1994 a total of 83 were counted on this bank, as against 6 on the much less warm east-facing bank on the opposite side of the road and 2 on the south-facing bank of Mill Lane which leaves the A538 westwards at this point. *O. venata* occurred higher up the slope where the vegetation was lusher, and in much lower numbers (only 3 were observed). The only other butterfly species seen on this visit was the Meadow Brown *Maniola jurtina* (14 on the west-facing slope, 3 on the east-facing, and 1 on Mill Lane); though the Narrow-Bordered Five-Spot Burnet moth *Zygaena lonicerae* was extremely abundant. This site stands to be destroyed in the construction of a second runway for the airport.

Mounds of soil, when grassed-over a few years after disturbance, also form good habitats. As an example, at Hunger Hill near Bolton (SD6806), on 8.vii.1995, large numbers of this species were again found co-occurring with lesser numbers of *O. venata* and with *Z. lonicerae* which abounded in myriads but was more localised. The vegetation where *T. sylvestris* was densest comprised sedge, bird's-foot trefoil and some *Holcus* grass; smaller numbers were on adjacent more recently disturbed ground dominated by buttercups, great hairy willow-herb and ragwort. In nearby longer grass with thistles, *O. venata* dominated.

The former Stretford tip in the Mersey Valley (SJ7893/4), now an extensive undulating grassland, has developed into ideal habitat.

Hostplants (Figs.102 & 148). Grasses, particularly Yorkshire Fog *Holcus lanatus*.

Broods. Single-brooded; the species has a short flight period, which starts in early July, or late June in favourable years. In 1994 it did not appear until 12.vii. Numbers usually build up very quickly to a peak when it is often very abundant, but then it declines very quickly and is over by early August. The flight is timed 2-3 weeks later than *O. venata* although there is an overlap so that the two species are seen together for a short period.

Distribution (figs.10, 37, 82 & 128). This butterfly has undergone a spectacular range extension and has colonised a large proportion of the mapped area since 1990. Prior to that year there had only been a handful of records, in the extreme south. In 1990, recorders in the Mersey Valley (SJ7893) and Heald Green (SJ8485) first noticed the butterfly; each year since it has spread further and increased in numbers. It now can confidently be expected in any suitable site south and west of the city; sites near the city centre have been colonised, an example being the disused Ardwick railway yard (SJ8697) where large numbers were observed on 20.vii.1995, and it continues to spread in the north.

It was first suggested that this butterfly might have dispersed by way of the railway network (Hardy, Hind & Dennis 1993). Also, as several of the first sites to be colonised were near motorways it is possible that some of the spread may have been along motorway verges.

Natural corridors such as river valleys also probably assisted. It is however clear that fecund females must have flown over very large areas of completely unsuitable terrain, both rural and urban, to locate some of the small, isolated patches of habitat (Dennis, 1992). In 1995 and 1996, when more parts of Sale were surveyed in greater detail, the species was found on virtually every marginal site checked including scraps of waste land in built-up areas.

Far fewer records were made in 1997. This may have been a result of recording bias (i.e. concentrating more on sites where this species is unlikely to occur), or it could be that the range extension is about to be followed by a contraction.

Behaviour. This butterfly builds up very large populations which give the impression of being confined to a small space, but in fact the species must be very mobile to have colonised so many new areas so quickly. Several individuals are often seen nectaring together, and they do not appear to be as aggressive to other butterflies as *Ochlodes venata*; they also fly and perch at a lower level. In most summers it has been apparent that *T. sylvestris* favoured sites which were noticeably warmer and more sheltered than surrounding land. In 1995, however, when June and early July were much hotter than usual, numbers in some of the previously favoured sites were much reduced, probably because the grass had become too desiccated; but the butterfly spread to other sites nearby. For example, although much less numerous on the slope beside the road north of the airport tunnel than in 1994, it was very abundant on some field edges about a mile to the north of this site (SJ8086); parallel instances were noted in the Mersey Valley where it was less numerous in the original sheltered site (SJ7893) but much more abundant elsewhere.

LARGE SKIPPER, *Ochlodes venata*, (Bremer & Grey, 1852)

Habitats. The species tolerates a wide range of habitats, including river valleys, grassy places in woodlands, scrub areas, railway verges, waste land, even sheltered spots on moorlands (for example, Holcombe Moor (SD7716), where it was noted at an altitude of 350m on 2.vii.1995). The main requirements are grassy areas with nectar sources plus taller vegetation for shelter and on which to perch. The habitat requirements often coincide with those of the Meadow Brown *Maniola jurtina* and very often the two species are found together, although their behaviour is very different. *O. venata* is not usually as abundant as *M. jurtina*, although if anything it tolerates a slightly greater range of biotopes. It also sometimes occurs along with the Small Skipper *Thymelicus sylvestris*; where this happens the present species is usually less numerous and prefers a different part of the site with more shelter and taller vegetation. For instance, in an observation on the warm, dry, west-facing slope beside the A538 road north of the tunnel under the airport runway (SJ8083) on 10.vii.1994, where there is a very strong colony of *T. sylvestris*, only 3 *O. venata* were recorded, and these were at the top of the slope where the vegetation was lusher and included some bushes.

Hostplants (figs.102 & 148). Grasses. The precise species used in the Manchester area have not been identified; possibly *Molinia* on mosses?

Broods. The single annual brood is from early June (late May in some years, for example 1992 when it was seen on 28.v) until early July. There is a short peak when the species may be very numerous, then a rapid decline and very few are seen after mid-July. The timing of the flight period is usually 2-3 weeks earlier than *T. sylvestris*.

Distribution (figs.11, 38, 83 & 129). Probably distributed over the whole area except the highest ground in the north and east. It has certainly been much longer established than *T. sylvestris*. Numbers are subject to much fluctuation from year to year, almost certainly due to weather conditions: it was very abundant in 1993 and 1995 and, apparently, scarce in 1996 and 1997.

Behaviour. This is a very active, territorial and aggressive species. It is fond of basking on bushes and other taller vegetation in corners or sheltered spots in hedgerows. Dennis (pers.

comm.) remarks that he has seen, and even heard, these butterflies strike each other in territorial disputes. They also successfully drive off much larger butterflies which invade their territories; examples noted during the survey have included a Comma *Polygonia c-album* (Partington (SJ7291), 21.vi.1995) and a Small Tortoiseshell *Aglais urticae* (Ladybrook Valley (SJ9085), 2.vii.1995). They have also been seen to chase off bees.

The few observations of this species during the 1996/7 nectar survey showed creeping thistle as the most popular nectar source, and tufted vetch second. From observations in other years, tufted vetch would seem to be the most widely used source, and in many of the butterfly's haunts in a good season almost every patch of the plant will have several feeding on it. Brambles are also popular and serve as perching sites as well as providing nutrition (Dennis & Williams, 1987). An unusual nectar source being used on the Mersey bank (SJ788929) on 7.vii.1994 was hedge bindweed.

Family PIERIDAE

CLOUDED YELLOW, *Colias croceus* (Geoffroy *in* Fourcroy, 1785)

Habitats. In the rare years when this migrant has reached the Manchester area, the river valleys, notably the Mersey and Bollin, have proved attractive to it. At the time of the butterfly's abundance in 1992, the weather conditions had provided ideal breeding conditions on the north bank of the river Mersey in Stretford, a section of my regularly surveyed route, with abundant hostplant, nectar sources (primarily ragwort and knapweed) and taller vegetation for shelter.

Hostplants (figs.110 & 156). Clovers. It will use them when they occur on almost bare ground (Dennis, pers. comm.)

Broods. From the observations in 1983 and 1992 it seems likely that the butterflies seen in July and August were British-bred descendants of immigrants which arrived unnoticed in May or June. In 1992, from the number of sightings along the Mersey bank (SJ7892/3) in July there is strong evidence that the species was breeding there; unfortunately there was a change in the weather, which following a warm spell became cold and wet at the end of July, and a mowing of the river bank by the water authority destroyed most of available hostplant and nectar sources and there were no further sightings along the bank that year. Reports from the Stockport area however suggested that breeding continued in suitable sites, including the revegetated Adswood tip (SJ8887), until October.

Distribution (figs.12, 40, 84 & 130). The majority of records are from the south of the area, including the Mersey Valley, Heald Green (SJ8485) and the Stockport and Wigan districts, and, sadly, reflect more where the recorders happened to be during the 1983 and 1992 immigrations rather than give any idea of the true extent of the butterfly's distribution in those years. There were also a few sightings in other years, especially in the Wigan area where there are a number of active recorders.

Behaviour. In the Mersey Valley (SJ7892/3), the butterflies have been seen actively patrolling to and fro along the berm on the warm south-facing slope of the north bank of the river. This is built up in the form of a levee for flood defence and has regularly served as a flyway for a number of patrolling butterflies; it was ideally suited to this species. The butterflies regularly paused for nectar - one incidentally was disturbed by a Gatekeeper *Pyronia tithonus* whilst nectaring from knapweed on 16.vii.1992, this being the first ever confirmed sighting of the latter species in this section of the Valley!

Even that controversial non-indigenous plant Himalayan balsam, which has taken over so much of the river valleys, had its place in this butterfly's ecology during the brief 1992 abundance, as on two separate occasions one was observed to select the plant as a resting-place, the butterfly's greenish-yellow underside and red antennae being remarkably well camouflaged when under the balsam leaf.

BRIMSTONE, *Gonepteryx rhamni* (Linnaeus, 1758)

Habitats. Most of the records are from the well-vegetated river valleys. Hedgerows or woodland edges or rides are essential.

Hostplants (figs.103 & 149). Alder Buckthorn *Frangula alnus*. This shrub has been planted in many sites in the area where it does not occur naturally, including in the Mersey Valley; this planting has clearly assisted the spread of the butterfly.

Broods. One single very long-lived brood per annum; butterflies are seen in late summer and early autumn (the latest recorded dates in this survey being in early October, in 1993 and 1995) and in April and May after hibernation. Butterflies seen in June may be either newly emerged, or overwintered ones still alive.

Distribution (figs.13, 42, 85 & 131). Prior to the early 1990s there were very few records of this butterfly in the mapped area and it was assumed to be an occasional wanderer from further south in Cheshire. Since then there has been a slight but noticeable increase, and, although it is still a rarity, it has been confirmed that the species breeds in parts of the Mersey Valley, including the Sale water-park (SJ7992), where five caterpillars were found on a mud-bespattered alder-buckthorn bush in the hedge beside the access road on the south side of the lake in July 1994. Breeding has also been reported from Urmston Meadows (SJ7693/4), and in artificially-planted "landscaping" woodland bordering open grass at the edge of the "Racecourse" local authority housing estate, in the west of Sale (SJ7691), where larvae were found in 1996. It is not yet possible to state whether the other records shown on the map, by other observers, represent established breeding populations or strays, though the species is known to breed in the Bollin Valley just outside the mapped area.

Behaviour. Sightings have mostly been of single butterflies, or at the most two. Often they are actively moving; most of the early records (1992/3) in the Mersey Valley were of individuals apparently on passage, flying purposefully in an east-west direction; more recently however they have seemed more settled and have been seen nectaring, including from buddleias.

LARGE WHITE, *Pieris brassicae* (Linnaeus, 1758)

Habitats. Woodland edges, river valleys, waste land and disturbed ground starting to revegetate are probably the best places for this species now. It does occur in agricultural land and suburban gardens but apparently not as much as formerly. It is highly mobile and individuals are often seen on passage flights seeking new habitats.

Hostplants (figs.104, 105, 150 & 151). Crucifers. As the species does not appear to be more abundant around the cabbage-fields at New Manor Farm and Mosley Acre Farm, Stretford (SJ7893, 7993), and the suburban allotments where cabbages and other Brassicas are grown, than in other areas the inference is that it breeds more on wild than cultivated hosts - or has a better survival rate on them. It has very different biology from the Small White *P. rapae* as although it utilises the same hostplant species it needs larger plants for optimum survival, as it lays its eggs in batches and the caterpillars are gregarious. The survival rate can however be very low due to parasitism; it is also possible that when the species attempts to breed in allotments the gardeners find and destroy the gregarious caterpillars more readily than the solitary larvae of *P. rapae* and thus reduce its breeding success compared with that species.

As an extreme example of this species's ability to seek out very small habitats amidst unfavourable environments, on 3.ix.1992 G. Bennion found a large brood of caterpillars on an isolated self-seeded Brassica plant with small, narrow leaves (species undetermined) growing at the edge of an alley behind terraced houses in Meadows Road, Sale (SJ7992). Perhaps even more remarkable is that not only had a female butterfly found the plant but also the butterfly's hymenopterous parasite had located it, as a good half of the caterpillars proved to be parasitised. The caterpillars were almost mature when discovered, but the plant had been virtually stripped and there was no other nearby, so whether any would have survived to pupation if they had not been brought into captivity is uncertain. The same year, Bennion also found an instance of batches of eggs of this butterfly and the Cabbage moth *Mamestra brassicae* laid on the same leaf of a wild Brassica plant.

Garden nasturtium *Tropaeolum majus* is also used as a host. Dennis (pers.comm.) remarks on the use of biased aspect by *P.brassicae* egg-laying on a circle of *T.majus* growing at the base of a garden buddleia.

Broods. Two broods per annum. The season is usually slightly shorter than that of the Small White *P. rapae* but slightly longer than the Green-Veined White *P. napi* with sightings from late April to September, a few stragglers lingering into October. The broods overlap, but there is normally a distinct peak in May and a much larger one in late July and August. In 1991 and 1997 the summer brood was noticeably longer and the abundance continued well into September.

Distribution (figs.14, 43, 86 & 132). The records show this species to be generally distributed, but not nearly as abundant in recent years as the Small and Green-Veined Whites. It also seems to be becoming more of a rural, and less of a suburban, butterfly. The Whites are always thought of as being very common; however that is partly an artifact of their apparency; being very conspicuous and mobile they seem more relatively abundant compared to other butterflies than they actually are. There was a wave of abundance in the summer of 1997, when numbers of this species approached half those of *P.rapae*.

Behaviour. The flight of this species is stronger than the other pierids; it tends to fly higher and ascend more quickly, frequently lifting straight up and over a line of tall trees with no apparent effort. It is normally fairly easy to distinguish it from the other Whites by its larger size and brighter white; however, small or worn individuals can be confused with large *P. rapae*. Although there are definite differences in the ecological requirements of the three *Pieris* species, there is considerable overlap and often two, or all three, are seen together. Sometimes interactions are of an aggressive nature; also frequently patrolling males of one species will investigate any other white butterfly, including a male of another species, in search of a potential mate.

In the 1996/7 detailed nectar survey, this butterfly showed a much greater propensity for buddleia than the other Whites. From previous years' observations it had been noted that it favoured thistles; during the 1996/7 survey, however, it was only seen using these on four days. Occasionally, unusual sources are used, including Himalayan balsam *Impatiens glandulifera* (Priory, Sale (SJ8092), 31.viii.1992, and once in 1997). The butterflies are seen far more often flying than nectaring.

SMALL WHITE, *Pieris rapae* (Linnaeus, 1758)

Habitats. This is a very successful species which can tolerate a wide range of habitats, and including those considerably modified by man. It frequents river valleys, woodland edges, waste ground especially in the early stages of regeneration where cruciferous plants are among the first to recolonise; also farm land with Brassica crops, and suburbs specifically where there are allotment gardens. These occur in a number of districts around Manchester, and often support large numbers of this species as Brassica crops are regularly grown in them. In the Winstanley Road allotments, Sale (SJ7992), during a half-hour visit on 11.vii.1995, 52 of this species were recorded as against 2 *P. napi*, 8 *Maniola jurtina*, 7 *Aglais urticae* and 4 *Thymelicus sylvestris*; during another visit to the same site on 22.viii.1995, 47 *P. rapae* were recorded with only 3 other butterflies, one each of *P. brassicae, A. urticae* and *Pararge aegeria*.

Hostplants (figs.104, 105, 150 & 151). Crucifers. Cultivated Brassica crops are used to a large extent, but the species certainly does also use wild crucifers, primarily the yellow-flowering wild Brassicas which quickly colonise disturbed ground. As with *P. brassicae*, the females have extraordinary powers of locating very small scraps of potential hostplant. On 4.viii.1994, one was seen ovipositing on pioneering crucifers only two or three inches high (shepherd's purse *Capsella bursa-pastoris*, an unidentified yellow species and at least one other) on a small patch of disturbed soil in an alley behind Harley Road, Sale (SJ7892). At that time the species was more than usually abundant in the Sale area. The hostplant-habitat maps possibly understate the true range of the potential wild hosts, as locating and mapping every small scrap of a plant such as shepherd's purse is practically impossible, and in doing this mapping I tended to concentrate on plotting the easier-to-see yellow-flowering crucifers. Oviposition on horseradish was observed twice in July 1997 in Sale, one occasion being in the Winstanley road allotments on 14.vii.97. Another occasional host is garden nasturtium *Tropaeolum majus*.

Broods. There are two main broods, but they overlap and the species is continuously on the wing from early April until September or October. The greatest abundance is in late July and August. As with *P. brassicae*, in 1991 the season was longer and there were many September sightings; judging by a sudden wave of abundance in the Mersey Valley on 8-9.ix.1991 there could well have been a third brood that year. There was also a late emergence in 1994, which may have represented a third brood, though it was not nearly as pronounced as that in 1991 and there were very few late *P. brassicae* or *P. napi* in 1994.

Although in the spring brood the numbers of this species and the Green-Veined White, *P. napi*, are nearer equal, this species usually produces a much stronger and earlier summer brood than *P. napi*. Possibly *P. rapae* pupae have a lower winter survival rate than *P. napi* owing to the preferred habitats being more susceptible to human disturbance as a result of crop harvesting, but *P. rapae* is better able to build numbers up during the summer.

Distribution (figs.15, 44, 87 & 133). This species is likely to be seen anywhere in the mapped area apart from some high ground. The shortage of records around Bury is due to lack of recorder coverage; one recorder who supplied a number of records for much of north Manchester for some reason did not record this species or *P. brassicae*.

P. rapae is normally the most abundant and most widely distributed of the three *Pieris* species; however, the possibility exists that it may have been slightly over-recorded because of the difficulty in distinguishing between it and *P. napi* (and sometimes, especially in the summer brood) in distinguishing large individuals from *P. brassicae*) in flight.

Behaviour. On warm days the butterflies are restless and many individuals move out of the breeding site seeking new habitats. Nectar stops are usually very brief; however, during the 1996/7 nectar survey this species used a greater variety of nectar sources (59 different plants) than any other. During the nectar survey thistles were the source with the highest number of records; dandelions were the second and brambles third. Very occasionally these butterflies take nutrition from other sources; they have been observed on wet mud on canal banks and on 15.viii.1995 one was seen feeding on dung.

Normally when these and other Whites bask, they only hold their wings three-quarters open. At the end of the season, however, lower temperatures may cause them to thermoregulate with wings seven-eighths open, aligned exactly to the azimuth and with the abdomen slightly raised; for example three were noted doing this by the Rochdale canal, Ancoats (SJ8598) on 23.ix.1994. They may have been signalling as well as basking? Dennis (1993) remarks that the species is probably a dorsal reflectance basker, but this mechanism is in doubt and it seems at least partially to involve dorsal (basal) absorption; the behaviour just described certainly seemed to constitute absorbence rather than reflectance basking.

The females' ability to locate very small, isolated patches of hostplant/habitat has already been remarked on. Similarly, males have extraordinary powers of locating females even in areas of very low density, as evidenced on 11.9.1994 when a mating pair were found sheltering from wind in long grass within a small thicket in the middle of a harvested cereal field at Dairyhouse Farm, near Altrincham (SJ7589); there was no obvious hostplant/habitat anywhere nearby and the area was virtually devoid of butterflies.

On 8-9.ix.1991, in a sudden short wave of abundance, large numbers of freshly-emerged males had gathered along a linear shelter-belt consisting of brambles and scrub beneath a line of low trees, running north-south at the eastern edge of fields north of Mosley Acre Farm, Stretford (SJ7993) in which Brassica crops had been grown that season. Hardly any females were observed and there were very few butterflies on the following days.

GREEN-VEINED WHITE, *Pieris napi* (Linnaeus, 1758)

Habitats. This species occupies a wide range of habitats, but not as wide as the last; it favours low-lying damp places, such as stream edges and wet meadows, often in similar situations to the Orange-Tip *Anthocharis cardamines*. It also frequents woodlands where it shares shady habitats with the Speckled Wood *Pararge aegeria* and often these are the only two butterfly species which will tolerate these conditions. Although in general it prefers damper, more wooded habitats than the Small White *P. rapae* there is overlap between the species. For instance, at wet seeps and watercress beds beside a woodland near Walker Fold, Bolton (SD6712) on 29.v.1995, archetypal *P. napi*/*A. cardamines* habitat, one individual proved on close inspection to be *P. rapae*; and on 20.viii.1995 at the Chadderton sewage farm (SD8904), a wasteland/wetland habitat with much tall grass and tall herbs, again both species occurred although *P. napi* greatly outnumbered *P. rapae*. Conversely, in typical *P. rapae* habitat such as suburban allotments, occasional *P. napi* occur; see the instance of the Winstanley Road allotments, Sale, mentioned in the account for *P. rapae*.

Hostplants (figs.105 & 151). Many cruciferous plants including cuckoo flower *Cardamine pratensis* and garlic mustard *Alliaria petiolata*. These are also the Orange Tip's preferred hosts but although the two species occur together they do not generally compete for resources; *A. cardamines* generally lays its eggs on pedicels of florets of mature plants, whereas *P. napi* will lay on any plant part, including leaves, usually of small plants (Dennis 1985a) and its eggs are very much more difficult to find than *A. cardamines*'s. *P. napi* also regularly uses other hostplants such as watercress *Nasturtium officinale*.

Broods. The first brood starts in April and continues until early June; the second is from July to August or early September. There is some overlap between late butterflies of the first brood and early ones of the second, but the broods are more distinct than in *P. rapae*. Unlike in *P. rapae* and *P. brassicae* where the summer brood is much more numerous, from my observations the numbers of *P. napi* in the two broods do not differ as much - i.e. *P.napi* is less numerous compared to the other two species in the summer brood than in the spring brood. This however does not appear to agree with data from the national butterfly monitoring scheme (Pollard, Hall & Bibby 1986) which cites an example from Carnforth Marsh, north Lancashire, in 1981 when numbers of *P.napi* in the summer brood were vastly greater than in the spring.
It is possible that some populations on the moorlands in the north and east may be univoltine.

Distribution (figs.16, 45, 88 & 134). This butterfly occurs over most of the area, including quite high up in the hills, and its distribution probably remains fairly constant. As it has different ecological requirements from the Small White *P. rapae* and is not quite as mobile as that species, it is less often seen away from the breeding habitat and in consequence seems less generally distributed. The perennial difficulty in trying to distinguish this butterfly from *P. rapae* in flight can cause considerable problems with recording. If anything *P. napi* has a slightly greyer look, sometimes almost bluish-grey, and usually does not fly as strongly. The chances are that a "small" White seen in a damp, sheltered habitat will be *P. napi* and one seen flying strongly through a suburban area will be *P. rapae*, but this is not always true and there are many times when the habitat, behaviour and appearance could fit either species. The summer brood can be even more difficult to distinguish as often the "green" veins on *P. napi* are very faint. Some of the records which I have received appear to show fewer *P. napi* in relation to *P. rapae* than I would have expected, and I fear that *P. napi* may have been slightly under-recorded.

Behaviour. As with other pierids, males patrol along linear faces such as woodland edges and river-banks in search of females; they have a slightly less vigorous, less sustained flight and seem slightly more inclined to settle or pause briefly to take nectar than *P. rapae*. In sites where the present species and the Orange Tip, *A. cardamines*, occur together, *P. napi*'s threshold temperature for flight is very slightly lower than *A. cardamines*'s, thus when the sun starts to break through on a dull day, as frequently happens in May, *P. napi* will start to fly slightly more readily than *A. cardamines*. The spring brood of *P. napi* continues on the wing slightly later in the season than the single brood of *A. cardamines*.

Although the 1996/7 nectar survey does not show it very markedly, in some sites great willow-herb *Epilobium hirsutum* is a very popular nectar source for the summer broods of all three *Pieris* species, but this one especially so as the plant frequently occurs in preferred *P. napi* habitat such as stream-sides and is in flower in late July and August. As well as nectar, these butterflies are very occasionally seen taking nutrition by "mud-puddling" in hot weather: observed instances during the survey were of single butterflies in the Tame Valley, Reddish (SJ9093) on 19.vii.1995, by the Manchester Ship Canal (SJ6990) on 23.vii.1995 and near Rochdale (SD9417) on 30.vii.1995.

ORANGE-TIP, *Anthocharis cardamines* (Linnaeus, 1758)

Habitats. The ideal habitats are stream and river valleys and low-lying damp meadows. In the hilly ground to the east the butterfly occurs up to 250 m altitude along stream edges. In the foothills of the West Pennine Moors, to the north-west, although the greatest concentration is to be found in valleys, it also occurs, at low density, in many semi-"improved" meadows on the lower slopes, in which there is an abundance of cuckoo-flower.

Some of this butterfly's habitats are in early stages of succession and without intervention by man stand to be lost. In the Mersey Valley, patrolling males often used to be seen along the edge of a recently-planted wood in Crossford sports field (SJ7993), and the butterfly likely bred in a south-facing "amphitheatre"-type clearing in the wood; the wood has now matured and the amphitheatre has become too overgrown and shaded. Further west in the valley at Newcroft tree nursery, Urmston (SJ7894), many young trees were planted in a marsh in the early 1990s. A strong growth of cuckoo flower established itself in the soil disturbed by the planting, and in 1995/6 the butterfly was breeding very well at this site; however, in a few years the hostplant will be shaded out as the trees grow. The butterfly is also regularly seen along the river: the males patrol the banks and the females oviposit on garlic mustard which grows low down on the banks. The banks are mown, sometimes two or three times per annum, by the water authority, who believe that the mowing increases the ability of the grass to develop strong roots to stabilise the bank; this mowing must destroy many ova and larvae every year but the butterfly persists. It is possible that some of the patrolling butterflies have bred in less disturbed parts of the valley and moved in to use the river-bank as a flyway.

Hostplants (figs.106, 107, 108, 152, 153 & 154). Normally cuckoo flower *Cardamine pratensis* and garlic mustard *Alliaria petiolata*. The first is the most favoured, and in some sites where it is not too dense every plant will have some eggs. With both hosts, plants in small clumps, or even single isolated plants, are far more likely to have eggs than individual plants in large expanses. In large open fields full of cuckoo flower fewer eggs than expected tend to be found, and most of them are concentrated around the periphery. This is the "density effect" as a result of flight behaviour, explained by Courtney (1982a, b) - the butterfly appears to aim for a spatially even distribution of eggs rather than a distribution proportionate to the abundance of the hostplant. For differences between "density" and "edge" and "recess" effects, see Porter *in* Dennis (1992), page 69; this paper illustrates an example from an "amphitheatre" in the Bollin Valley near Rossmill (SJ7884).

As this is quite the easiest of any butterfly species to find in the egg stage, and searching for eggs can be very rewarding when the weather is too dull for flight, a number of checks were made in 1995 for possible alternative hosts. Eggs were found on sweet rocket (=dame's violet) *Hesperis matronalis* (a few in the "Priory" in the Mersey Valley, Sale (SJ7992), and a number on an isolated clump of the plant in the Bollin Valley south of Bowdon (SJ7585)), honesty *Lunaria annua* (Marple area), common wintercress *Barbarea vulgaris*, watercress *Nasturtium officinale* (Mersey Valley, Ashton-on-Mersey (SJ7792)), wild turnip *Brassica rapa* (Douglas Valley, Gathurst (SD5208)) and an "unidentified wild Brassica amidst lots of garlic mustard" (Roch Valley near Hooley Bridge (SD8411)). It appeared that these alternative hosts were being selected later in the flight period. Possibly the females recognise that by this time the cuckoo flower and garlic mustard have become too senescent for the larvae and therefore elect for a less suitable plant in an earlier stage of flowering. It is however also possible that recorders have become so used to searching the usually accepted hostplants that they do not give enough attention to alternatives, and therefore many eggs on other plants go unnoticed.

Courtney (1980) investigated whether larvae were able to complete development on these alternative plants, and suggested that they have varying success. Much research remains to be done; the only relevant observation made during the present survey is that half-grown larvae were found on wild turnip in the Douglas Valley, near Gathurst (SD5208) on 14.vi.1995.

On 11.v.1996, a female butterfly was observed ovipositing on fool's parsley *Aethusa cynapium* at Crossford sports field, Sale (SJ7993).

Broods. Unlike the other white pierids this butterfly has only one brood. Emergence normally starts in late April, sometimes earlier in the month as in 1997 when the first was seen on 8.iv; the peak is usually over by the last week in May, and sightings after the first week in June are rare. Females appear to linger on longer than males. The flight period is slightly, but noticeably, later in the north of the area than in the generally lusher south.

Distribution (figs.17, 46, 89 & 135). This butterfly is probably a fairly recent colonist in much of the area, though the date of colonisation is unknown. In the early 1970s I saw it near Woodford (SJ8981) in the extreme south-east, but it did not then appear to occur in the Mersey Valley. Now it ranges throughout most of the area, having spread through all the river valleys, as well as along man-made flyways such as canals and possibly railways, virtually to the city centre, and outwards and upwards along streams into the Pennine foothills. Females clearly must move about a great deal as frequently eggs are found in the most unlikely-looking places where there have been no sightings of the adult. Anywhere where one of the hostplants has managed to colonise, the butterfly is likely to follow.

Behaviour. Males are frequently seen patrolling along defined flyways, such as river banks and woodland edges. Males usually seem to outnumber females by about 10 to 1; in reality the numbers of each sex are about equal but the males are much more inclined to fly than the females and are therefore much more apparent; also the females are easily mistaken for other white pierids.

Sometimes it is said that females do not usually lay on plants where there is already an egg (Courtney, 1981 and Thomas, 1984, but see Dennis, 1982). In the present survey this certainly has not always been found to be the case. Many times plants have been found with two or more eggs, and once, in the Mersey Valley near Northenden (SJ8290) a single plant with nine eggs; in that case there were few other plants nearby and it would seem that several females must have been drawn by the same cues to what appeared to be an ideal plant. Sometimes also newly-laid eggs and half-grown caterpillars have been found together, and several times more than one caterpillar on the same plant, in spite of their tendency to cannibalism.

Normally females select plants for egg-laying which are growing in fairly open sites not too far from a shelter-belt; plants growing in shade are generally unsuitable. The present survey has proved however that there are exceptions to this also, as for example an egg was found on 21.v.1995 on an isolated plant of garlic mustard in heavy shade under a Manchester poplar tree in a wood at the confluence of the Glaze brook and the Manchester ship canal, Irlam (SJ7091).

Family LYCAENIDAE

GREEN HAIRSTREAK, *Callophrys rubi* (Linnaeus, 1758)

Habitats. Most of the few known colonies are in the Pennine foothills, on south-west facing slopes or hollows which provide a much warmer microclimate than the surrounding open hills. Examples are Dick Clough, Tunstead (SE0004), where a stream runs through a steep valley with abundant semi-natural vegetation including plenty of gorse and bilberry and also open woodland, providing a warm, sheltered site; a location near Brookbottom (SJ9886) where gorse and bilberry grow on a steep south-facing hillside; and Cock Wood in the Brushes Valley (SJ9998) near Stalybridge, at the bottom of a slope which although north-facing is warm and sheltered because it is bounded on the north side by the edge of a wood. Large expanses of heather and bilberry on open hillsides are not suitable habitat.

Just outside the recording area, the butterfly has been found in a different type of habitat: low-lying peat bog at Lindow Moss near Wilmslow.

Hostplants. It is believed that all the colonies in this area use bilberry *Vaccinium myrtillus*. As gorse *Ulex europaeus* is present in some of the sites it is a possible alternative.

Broods. Single-brooded; it should be looked for in May, or possibly from late April, as colonies further south in the Pennines (near Edale) are often out by then.

Distribution (figs.18 & 47). Most of the known colonies are in the extreme east of the recording area and only just within the Greater Manchester boundary. In addition to those localities mentioned under "Habitats" it has been reported from Werneth Low, Hyde (SJ9693). It also just penetrates the north-west of the area, having been recorded at Foxholes, north-west of Horwich (SD6412) in 1993. There are a number of colonies further to the north-west of Bolton, outside the area; S.P.Garland (pers. comm.) believes that the butterfly may be extending its range there. There are a number of likely sites around Bury and Rochdale, between the two known centres of distribution, and although my own searches in 1995 and 1996 were without success, I have received a report via the Bolton museum that the butterfly was found near Holcombe Brook, north of Bury (SD7616/7) in 1996 by C.Johnson. There would not seem to be much opportunity for range extension in the eastern sector as there are only isolated pockets of suitable habitat; rather, the colonies would appear to be the relicts of a former extensive distribution when the hill vegetation was less depauperate.

Behaviour. Although not very conspicuous, this is by far the easiest to find of the three Hairstreaks occurring in the area, as its hosts are low-growing and it flies at eye-level. Normally when one butterfly is located, others will readily be found nearby. They give the impression that they are unlikely to fly far and that there is unlikely to be much if any movement between colonies. Like some other apparently sedentary butterflies, however, they may be more mobile than imagined; C.I.Rutherford of Alderley Edge, Cheshire (a few miles to the south of the mapped area) has recorded one in his garden well away from the nearest known habitats at Lindow, Wilmslow. Even more surprising is a report from 1987 by N.W.Catchpole, an experienced butterfly watcher and photographer, of a sighting in Dark Lane nature reserve, Carrington Moss (SJ7390). There has not been any other report, and neither is there any area of apparently typical habitat, anywhere near.

PURPLE HAIRSTREAK, *Quercusia quercus* (Linnaeus, 1758)

Habitats. Normally woodlands with mature oak trees.

Hostplants. Oaks *Quercus robur* and *Q. petraea*.

Broods. One annual brood, in July and early August.

Distribution (figs.19 & 48). The only confirmed sightings within the area are several reports from a site at Billinge (SD5202), and an extraordinary record from Hale (SJ7886), a district of large suburban gardens, on 2.viii.1995 (K.R.C.Neal via R.L.H.Dennis). On 11.viii.1995 there was an unconfirmed report from Carrington Moss, beside the disused railway near the Altrincham sewage farm (SJ7590). There were several large oak trees nearby and the site looked suitable for the species. Many other locations in the south of the area appear suitable, and the butterfly has been reported from the Bollin Valley, and in a garden in Alderley Edge, both just outside the area. It almost certainly will occur in other sites but be overlooked.

Behaviour. This butterfly normally keeps to the tops of trees and is rarely seen. Occasionally it descends to take nectar; the one at Hale was photographed nectaring on a garden buddleia.

WHITE-LETTER HAIRSTREAK, *Satyrium w-album* (Knoch, 1782)

Habitats. The ideal habitat is the edges of mature woods containing elms, though it will breed on an isolated tree.

Hostplants. Elms. All records in this area are of wych elm *Ulmus glabra*.

Broods. One generation per year. The best time to look for it in this area appears to be late July and the beginning of August, although further south in Britain the beginning of July is the optimum time.

Distribution (figs.20 & 49). The historic records imply that this butterfly was formerly much more widely distributed in the area than it is currently; this is no doubt largely a result of loss of its hostplant through disease. A now well-known colony was discovered at Chadkirk (SJ9389/9489) by A.R.Blears some years ago. Up to 1986 there were still several large, mature wych-elms at the edge of a wood by the picnic site bordering the river, and the butterfly bred on them. During the next few years all those elms died. Some sucker growth remained by the stumps of felled diseased elms beside the approach road, and one mature tree near the car park remained. The butterfly survived in reduced numbers mainly on this one tree. In 1995 the tree was starting to show signs of disease; the butterfly was still present on it. It was also found at another site nearby, on a small, healthy elm on the side of a sunlit glade on the river bank. There have been reports from the Etherow country park, Compstall (SJ9790), and it would probably be worth searching any elms in the south and east of the area. In 1996 R.S.Greenwood recorded one at Plumpton Top, Heywood (SD8511), in the Roch Valley, an area with no previous reports.

There was also a report, by A.J.Thorley via the Oldham museum, of a claimed sighting on Harrop Edge, Diggle (SD9908) on an unspecified date in 1990; the recorder, a policeman, said that a butterfly settled on his "Land-Rover" while it was parked on a moorland track and he felt 100% confident of its identification having previously been involved in a project breeding the species. The locality is open windswept moorland and there are no elms (or any other trees) nearby. If correct, the butterfly must have been a stray carried a very considerable distance off course. I have not included the record on the distribution map.

Behaviour. Up to 1986, when there were elms at the wood edge at Chadkirk, the butterflies would come down to nectar on bramble flowers on the steep south-west-facing slope at the edge of the wood. In more recent years the only sightings at this locality have been of butterflies flying around or walking over the branches of the host trees. The butterfly seen by R.S.Greenwood at Plumpton was nectaring on creeping thistle. From leaf damage at the top of the elms in and near the known breeding areas, it is possible that this secretive butterfly could be a lot more numerous than the records indicate.

SMALL COPPER, *Lycaena phlaeas* (Linnaeus, 1761)

Habitats. This butterfly favours areas of rough grass mixed with flowers plus some taller vegetation to provide shelter and territorial perches. The habitat needs to be fairly warm. Suitable sites occur in many biotopes including river valleys, reclaimed tips, waste ground and regenerating sites. A certain amount of light mowing can be tolerated, as on the south-facing slope of the Mersey bank, Stretford (SJ7892); this bank usually receives one or two cuts per annum, which keep the grass fairly short without destroying the hostplants. The butterfly has also been found in the less mown parts of some cemeteries (Manchester Southern (SJ8292) and Weaste (SJ7997)). Very often it shares the habitat with the Wall *Lasiommata megera*. Level hay meadows, of which there are several in the Mersey Valley, although they contain abundant sorrel, are not suitable habitat.

Hostplants (figs.109 & 155). Usually common sorrel *Rumex acetosa*, but it has been recorded using dock. Development on dock will be slower and the larvae more vulnerable. The 7 X 5 km and 3 X 2 km maps show sorrel only.

Broods. Two or three broods per annum. The first brood is out in May (early or late May depending on how advanced the spring is) to June, the second in July and August, and the partial third brood in September, occasionally October; in 1997 October was the month with the most sightings in the Mersey Valley and they continued until 27.x. There is much overlap between late examples of one brood and early examples of the next and in different sites broods may be differently timed; a third brood may occur in some sites but not others. The second brood is normally much more numerous than the first. For example, in 1995 I recorded only three first-brood butterflies anywhere and none in the Mersey Valley; the second brood was much better, although it had nothing like the abundance of some years, 1989 for instance. Probably as a result of the unusual heat of 1995's later summer, there was an unusually strong third brood in some localities; in mid-September there were exceptional numbers in the Bollin Valley, Wilmslow (SJ8581, outside the recording area); a strong late emergence was also noted in a colony partly in Weaste cemetery (SJ7997) and partly on waste land just outside.
As well as the habitat coincidence mentioned above, the broods of this butterfly often appear to emerge in close synchrony with the Wall *Lasiommata megera*. The environment would seem to suit these two quite unrelated species equally. Perhaps it may be significant here that they both have a colour scheme comprising mainly orange and dark brown.

Distribution (figs.21, 50, 90 & 136). This butterfly appears to be generally distributed but in varying numbers from year to year. Although it has generally been regarded as a sedentary butterfly it clearly does have considerable powers of dispersal and readily colonises new sites, for instance regenerating waste ground in Salford (SJ8297). In 1992, a single caterpillar of this species was found by chance when picking a dock leaf in the very small front garden of a terraced house in Dudley Road, Sale (SJ791926). This meant that although the spring brood had appeared to be very small and had gone over very quickly, a fecund female must have strayed at least half a mile from the normal habitat in the Mersey Valley and found this hostplant. Butterflies of this species are unlikely to be noticed when "on passage"; owing to their small size and fairly dark colouration they are not easy to see apart from when basking with wings open or feeding at flowers. Dennis & Shreeve (1996) have shown that many butterfly species formerly believed to be sedentary ("closed populations") are in fact not so.

Behaviour. Ragwort *Senecio jacobaea* is very popular as a nectar source. Other composite flowers are used, including, in the third brood, Michaelmas Daisy *Aster novi-belgii*. On 5.ix.1994 a butterfly was seen attempting to feed from stale animal dung on a south-west-facing slope by Great Clowes Street, Salford (SD8200).
In 1989, the species was exceptionally abundant in the Mersey Valley, and it was possible to make some observations on its mobility. As an example of what appeared to be the extremely sedentary behaviour formerly believed to be characteristic of the species, I found an example of the form *"radiata"*, with the orange band on the upperside hindwing reduced to narrow rays, on the river bank close to the Stretford sewage farm (SJ7892) at 12.45 p.m. on 28.vii, then again at the same time on 2.viii in the identical spot, when I photographed it. It could of

course have been a different individual, as on returning to the spot 40 minutes after the first sighting on 28.vii (with a camera) I could not find the butterfly, and neither could I when I next visited the site on 31.vii.; however, having looked at it very closely each time I believe it was the same one, and that on the two intervening occasions when I did not see it this was because it was resting nearby but out of sight.

On 27.ix I photographed another *radiata* in the same place. These were my first sightings of this form in the area for thirty years; I had recorded one approximately half a mile from this site in 1959; the habitat was however very different then and the Stretford tip now covers the site where the 1959 specimen was found.

Following the 1989 abundance I found a number of eggs on sorrel on the Mersey river bank close to where the above observations were made, and brought them indoors in October and reared the caterpillars through the winter. The butterflies emerged at the end of April and I released them at the spot where the eggs had been found. Following release, each butterfly flew rapidly away, some heading along the bank but most going up and over the levee, and although I made frequent searches of the river-bank during the ensuing days I only saw one individual again. This could have been due to any of three causes: (1) dispersal of individuals: freshly-emerged butterflies may naturally have more of a tendency to leave the breeding site on their first flights; (2) disorientation of the released butterflies causing them to behave atypically; or (3) non-apparency - the butterflies, like the *radiata* the previous summer, may have remained in the area but simply not have been seen - the fact that one individual *was* seen again proves that at least some remain; it is reasonably certain that the individual seen again was one of the butterflies reared indoors as it was another two weeks before any "wild" examples started to appear.

LONG-TAILED BLUE, *Lampides boeticus*, Linnaeus, 1767

A single sighting of this rare migrant was reported to the Bolton museum by Mark Lightowler, manager of the Bolton butterfly house. The butterfly was seen on 1.viii.1995 in Queen's Park (SD7009), close to the butterfly house (fig.22). Queen's Park consists of an area of formal parkland laid out with lawns, planted trees and flowers, mainly non-native but with some indigenous vegetation along the banks of the river Croal which flows through it.

COMMON BLUE, *Polyommatus icarus*, Rottemburg, 1775

Habitats. Most colonies are on waste land, especially where there is a limestone content. Although limestone is not part of the natural geology of lowland Manchester, many habitats suitable for this species have developed from tipping or demolition sites which often include limestone from building foundations or roadstone, or from ballast along former railway lines on in abandoned railway yards. Some reclaimed tips, for example the fly-ash tip at Flixton (SJ7393), were artificially coated with lime-rich topsoil imported from the Peak District: this introduced a limestone flora and the butterfly colonised. In the west, many large populations occur on habitats associated with former coal-mining. There are some colonies in the "mosses" to the west, though they tend to be on roadsides or river-banks, or sites where there has been some introduction of lime. South-facing slopes are often favoured. In all habitats some shelter-belts, normally consisting of scrub, are present. The species often occurs with the Small Copper *Lycaena phlaeas*, notably on demolition sites which have been left alone for a few years but not yet become too scrubbed-over, although it uses totally different hostplants; it is less generally distributed but usually more numerous that *L. phlaeas* where it occurs.

Hostplants (figs.110, 111, 156 & 157). In habitats associated with railways, bird's-foot trefoil *Lotus corniculatus* is the main host. In these sites the plant tends to form dense low mats and is not out-competed by grasses and other vegetation. On many other sites, including former tips, river banks and demolition sites, the usual host is white clover *Trifolium repens*. Lesser trefoil *Trifolium dubium* is probably sometimes used. The large clumps of greater bird's-foot trefoil

(*L. uliginosus*) which grow in long grass in some areas, notably the Mersey Valley, are unsuitable and do not support the butterfly; neither do the vigorous clover plants which rapidly spread over former tips and other disturbed ground in the early stages of restoration. The butterfly's breeding success at any particular hostplant/habitat site may well depend on the presence of ants. See Fiedler (1990) and Fiedler & Holldobler (1992), which deal with the effect of larval diet on the attractiveness of the caterpillars of this species to ants.

Broods. The flight period is from late May to September (rarely October - in 1991 the latest recorded date was 6.x). The broods are ill-defined and seem to peak at different times in different sites. Although August is usually the month of greatest abundance, there is sometimes a high peak in June.

Distribution (figs.23, 54, 91, 137). This species ranges over most of the area, except the north-east, but is localised, occurring in small isolated sites amidst large expanses of unsuitable habitat. There are, or have been, a number of sites within a mile radius of the city centre, including the disused Pomona Docks (SJ8196), and railway sidings near Castlefield (SJ8397). Abandonment of industry and mining has clearly been a major factor in the spread of this species. Many colonies have become established on former industrial land in the west and north-west, including Horwich (SD6310) where scrub has developed on former railway sidings, the former tip at Nob End south of Bolton (SD7406) where dumping of oil of vitriol has created a very lime-rich, alkaline soil, Pennington Flash (SJ6499) where a very strong population thrives on a south-facing slope below the canal, the Wigan Flashes (SD5802/3) and a site between the disused railway and the B.P. oil works at Partington (SJ7291).

It is uncertain how recently these sites were colonised, as apart from in the Mersey Valley few observations are available from the 1980s. In the Valley, the butterfly seems to have been first recorded in 1983; then in 1984 there was an unprecedented abundance and in early August the butterflies were seen in numbers all through the Valley. The following year, 1985, their numbers were very much less; there were very few sightings in the Valley during the next few years, followed by a recovery in 1990. Since then, the species seems to have become stabilised. In the Valley it now has established colonies on the edge of the Stretford tip (SJ7893/4), the former Bethell's tip near the "Priory", Sale (SJ7992), and a limestone "raft" adjacent to the electricity substation near Kenworthy (SJ8191). Sightings on the Mersey bank (SJ7892/3) have become sporadic; probably the regular mowing of the bank has hindered the establishment of a permanent population. Further south, there is a good colony on the bank of the Sinderland brook, near the Altrincham incinerator (SJ7590); like the Mersey this river has built-up banks which form a warm south-facing slope. Although the Mersey Valley wardens have suggested that the 1984 abundance was a result of their planting bird's-foot trefoil at several sites in the Valley, this is extremely unlikely. Similar abundance that year was reported in Cheshire (Dennis 1985c) and it is likely that it was general, and almost certainly due to climatic conditions. Possibly a similar pattern of colonisation and fluctuation to that in the Mersey Valley has occurred over most of the area.

The scarcity or absence in the north-east of the map is probably genuine rather than a result of recorder bias. The hilly ground does not offer suitable habitat and also searches of several likely-looking sites in the Oldham and Rochdale areas did not find the butterfly.

Behaviour. Although most of the Blues have generally been regarded as sedentary species, *P. icarus* clearly does have a great deal of dispersal ability. As well as established colonies, several records are of single sightings. These could represent small colonies, but more likely are strays wandering out of the breeding area or dispersing, as often they are seen in habitat which does not appear suitable, yet not near any known colonies. Usually they are males; the females, being more cryptically coloured, are less apparent but clearly must disperse as much as the males.

There is much variability in the colours of the females, and blue and brown forms occur together. On 21.ix.1994 a single fresh female of an extreme blue form, with only the slightest trace of orange marginal lunules on the upper surface, was seen at Clayton Vale (SJ8899) in the Medlock Valley, on a south-facing slope abounding in clovers and bird's-foot trefoil.

Several apparently aggressive interactions have been observed during this survey between this species and others, including the Meadow Brown *Maniola jurtina*. A notable instance was

25.vi.95 at the edge of a former tip at Land Gate, near Wigan (SD5701), where a very small site with an optimum microclimate had become crowded with butterflies resulting in competition for space.

HOLLY BLUE, *Celastrina argiolus* (Linnaeus, 1758)

Habitats. Most of the sightings have been in gardens and parks and unlike every other butterfly species occurring in Greater Manchester this one finds more suitable habitat in suburbs, specifically those with large, long-established gardens, than in rural areas. There have however been some records from open countryside including the river valleys, usually of butterflies flying along hedges and in scrub areas with bushes and small trees.

Hostplants (figs.112, 113, 158 & 159). The only definite records are holly *Ilex aquifolium* and ivy *Hedera helix*. Females have been seen ovipositing on holly in Sale, and S.H.Hind's records prove widespread use of ivy by second-brood females in the Stockport area. As some of the sightings of adult butterflies have been in sites where neither of these plants grow, it is likely that other hosts are used also.

Broods. Although in some parts of northern England and the north Midlands (for example, Gaitbarrows, Lancashire and Coombes Valley, Staffordshire) this species has established univoltine populations, in the Manchester area it shows the normal bivoltine cycle with butterflies seen in May (sometimes from late April) and again in July to August.

Distribution (figs.24, 55, 92 & 138). Prior to 1990/1 the species had been very scarce in the Manchester area. In that period it underwent a population explosion over much of Britain, mainly in the south, but some of the explosion reached Greater Manchester. The map implies that this occurred especially in the Stockport area (the south-eastern sector); however, the concentration of records in this region is primarily due to S.H.Hind's very intensive searches for ova during the period of abundance, and therefore may not convey a true picture of the distribution. Hind believed it to be a range-expansion analogous to those of the Small Skipper *Thymelicus sylvestris*, the Gatekeeper *Pyronia tithonus* and the Speckled Wood *Pararge aegeria*. The lack of subsequent records from his area, and the general opinion that the species had passed the peak of its population explosion in 1992 and reverted to its normal frequency by 1993 indicated that, although *C. argiolus* began to spread at the same time as the above three grass-feeding species, unlike them it retained only a temporary hold in marginal habitats.

The situation in the Mersey Valley, however, seems different. First, my observations suggest that the influx appeared later and in less intensity than in Hind's area. *C. argiolus* was almost unknown in the Valley until 1991, when there was a single sighting, followed by a few more in 1992, including one found by G.Bennion in a Sale garden (SJ7992) on 21.vii. In 1993, the butterfly appeared at a number of Sale sites, but still in low numbers, in early May. Breeding was proved when on 6.v a female was seen laying on holly in the Sale golf club car park (SJ8091), a site where there are both holly and ivy and which for nearly forty years had struck me as potentially suitable for the species, but I had never previously seen it there. Others were flying around hollies in a garden on Dane Road, Sale (SJ7992), the previous day, 5.v., implying likely breeding. The butterfly was still present in the Valley in 1994 and 1995, when again it was seen at the Dane Road site, though its numbers remained low. Then in 1996 there was a distinct increase and in August single individuals were seen in several gardens and parks in Sale as well as Dane Road again. In 1997 it was notably more numerous and recorded from thirteen 100 m squares within the 7 X 5 km zone, as against a maximum of four in any previous year (see the table on page 47). Even so, the 7 X 5 km map probably grossly understates the true range of the species in Sale and it almost certainly would have been found in many other parts where hollies in gardens are more numerous.

During my recording walks elsewhere in Greater Manchester I have only twice come across this butterfly: in 1991 near Broadbottom (SJ9893) and in 1995 at Chadkirk (SJ9389), both these sites being within Hind's recording area. The lack of records from that area since 1991 may be entirely due to Hind's concentration on recording elsewhere and the butterfly may well have persisted at low density. To map this species accurately would require intensive recording by a team of recorders spaced throughout Greater Manchester.

Behaviour. Almost all sightings have been of single butterflies flying very actively round small trees and tall shrubs, very occasionally pausing briefly to nectar. The sightings are not usually followed by others on the same of subsequent days and the inference is that the butterflies must be constantly on the move over a wide area and able to exist at a very low density. Much remains to be discovered about how they locate mates and where their centres of breeding are.

Family NYMPHALIDAE (sub-family NYMPHALINAE)

RED ADMIRAL, *Vanessa atalanta* (Linnaeus, 1758)

Habitats. This migrant's foremost requirement is breeding habitat with nettles. As shown by the 7 X 5 km and 3 X 2 km maps, however, it is far more readily seen in nectaring than in breeding sites; it is the most extreme example of a species which moves out of its breeding habitat in search of nectar. All the large nymphalines do this to some extent, but this one particularly so, and because of its ready apparency to a casual observer it can give the uninitiated the false impression that it is most at home in flower gardens (Sanders, 1939) and that gardeners are doing butterflies a favour by clearing areas of "weeds" and planting gardens-full of flowers especially buddleias. This of course is not the full picture and "wild" sites are essential.

The ideal habitat for the species would appear to be woodland edges and rides. A single tree in a wood may form a territorial site for several weeks (Dennis, pers. comm.). The butterfly likes sunny gaps with some shelter from trees and sometimes shares a habitat with the Speckled Wood *Pararge aegeria*; its range of tolerance is however much greater. Following arrival in June the immigrants are often seen ranging through the low-lying "mosses" in the west, where there is a fair amount of this type of habitat. The locally-bred butterflies, in late summer and autumn, will home in to nectar sources, especially buddleias *Buddleja davidii*, which they appear to be able to smell from a considerable distance and in the most unlikely places: for example, on 24.ix.1994 three were seen on a bush close to Manchester Cathedral (SJ8398), there being no suitable breeding habitat anywhere in the vicinity. Self-seeded buddleia bushes have colonised waste land, and become a distinct feature of Manchester's flora, notably around railways and former railway yards; these are very popular with this species, as well as the many grown in suburban gardens. Buddleias have also been planted in new industrial areas such as Salford Quays (SJ8097) and these also draw the butterfly. In some years it favours the abundant Michaelmas Daisies *Aster novi-belgii* which grow in masses in regenerated sites often on former tips, including in the Mersey Valley (Kenworthy (SJ8191/8291), Stretford tip (SJ7893/4)) or the Croal/Irwell valley (Moses Gate country park (SD7307, 7407)).

Hostplants (figs.114 & 160). Nettles *Urtica dioica* and *U. urens*. Eggs are laid singly, and often on small clumps of nettles, unlike the Small Tortoiseshell *Aglais urticae* and Peacock *Inachis io*. Oviposition has been observed on a small scrap of nettle on the bank of the river Mersey, Stretford (SJ7893); and on a small patch of overgrown waste land in the middle of a housing area in Salford (Fitzwarren Street) (SJ8098).

Distribution (figs.25, 56, 93 & 139). In seasons of abundance this species is likely to be encountered anywhere in the area where there are larval or adult resources. Individuals are often seen on purposeful cross-country flights. The butterfly will cross most terrain without the need for defined flyways, although it certainly does sometimes use such; it has been seen using the Mersey bank. It will however balk at flying under wide bridges; an instance of this was observed on 12.ix.1995 when one heading up the line of a mineral railway in a deep cutting near Little Bolton (SJ7998) refused to fly under the bridge carrying Eccles New Road over the cutting. Being a migrant, the species varies greatly in numbers from year to year; 1995 was a year of exceptional abundance.

Behaviour. In early summer the immigrant butterflies are sometimes seen thermoregulating on road surfaces where there is not much traffic, as in the mosses. The thermoregulation behaviour is noticeably different from that of the Painted Lady *V. cardui*; although both species require warmth the Red Admiral's requirements are not so extreme; and frequently, when resting or feeding, *V. atalanta* will only hold its wings three-quarters open whereas *V. cardui* will hold them fully open. A corollary of this is that it is much more difficult to obtain a satisfactory photograph of the dorsal surface of *V. atalanta* than of *V. cardui*.

As an instance of the different thermal requirements of this species from *V. cardui*, one of each species was observed in the late afternoon of 11.ix.1994 at a small grassland site with some trees north of Sinderland Road, Broadheath (SJ7590), in hazy sunshine with clouds and moderate wind. At 4.23 p.m. the Red Admiral, with its lower temperature threshold for activity, was flushed, flew round a high hawthorn hedge, in circles, descended, basked on dumped dead privet with wings fully open, then was flushed again, flew into a high hawthorn tree and appeared to take roost in it.

Although in the "Indian summer" autumns of 1989, 1990 and 1991, in each of which *V. atalanta* was abundant, it very much favoured Michaelmas Daisies as a nectar source, in the following years its preference for buddleia above all other potential sources was very noticeable. On several days in 1992 butterflies were seen on a buddleia bush in a small garden of a terraced house in Sale (SJ791926) just outside the Mersey Valley, but not elsewhere on walks through the Valley. There was no breeding habitat near the house; it would be interesting to know how far the butterflies had come, whether they were the same individuals each day, and if so where they spent the nights. A recent paper by Wiltshire (1997) discusses mating, territories and roosting in this butterfly.

During the 1996/7 nectar survey, 77% of *V. atalanta* records/days were on buddleia; the species was only seen on 6 other plants. The other nymphalines also use buddleia but not to such a disproportionate extent as *V. atalanta*. As an example, on 2.ix.1995 on a patch of self-seeded buddleias on waste ground near Leigh (SD6500) there were 29 *V. atalanta* as against 3 *V. cardui*, 9 *Aglais urticae* and 1 *Pieris napi*.

Other non-indigenous flowers used by this species during the 1995 abundance were everlasting flower, African marigold (28.viii.1995), dahlia (16.ix.1995) iceplant (*Sedum*) (16.ix.1995 - seven were seen on this plant in a rural garden near Standish (SD5509), when none were seen in adjacent semi-natural habitat). On the other hand, that season there were a few sightings using native flowers, including several on hemp agrimony *Eupatorium cannabinum*, and on fleabane *Pulicaria dysenterica* and round-headed rampion *Phyteuma orbiculare* on the sunny south side of Ringley Wood (SD7605) in the late afternoon of 27.viii.1995, along with several Commas *Polygonia c-album*.

The only records during the survey of *V. atalanta* obtaining nutrition other than from flowers were near Wigan (SD5409) on 16.ix.1995 when some were seen on fallen apples, and 2.ix.1995 when four had congregated to feed on dumped rubbish in Lilford Park, Leigh (SD6600).

The species continues to fly later in the year than any other butterfly and in 1994 and 1995 there were sightings in November.

PAINTED LADY, *Vanessa cardui* (Linnaeus, 1758)

Habitats. The ideal breeding habitat is rough ground with abundant thistles. Warm, open areas such as the mosses in the west are popular. Being highly migratory, the butterflies are constantly investigating new areas as potential breeding sites or nectar sources, and may turn up almost anywhere.

Hostplants (figs.115 & 161). The only hostplant noted during the survey was creeping thistle *Cirsium arvense*; it is likely that spear thistle *C. vulgare* is also used. These plants are widely distributed, on waste land and also in many rural areas, where they occur in extensive patches on the nutrient-rich land in fields which have been abandoned or temporarily left fallow;

currently the butterfly has no shortage of potential breeding sites. In the Mersey Valley, some fields at New Manor Farm, Stretford (SJ7893), normally used for Brassica crops, were "set-aside" during 1996 and fortuitously a vast bed of thistles developed in them at the time of the butterfly's exceptional abundance that summer and it readily bred there. The farmer refrained from removing the thistles until the end of the season. The species was not this fortunate everywhere; for instance A.R.Blears drew the Stockport local authority's attention to the complete destruction of a large area of hostplant/habitat on a farm at Chadkirk (SJ9389) maintained by them as an environmental showpiece.

Distribution (figs.26, 57, 94 & 140). In most years this migrant species is generally distributed throughout the area, though its numbers vary greatly. Normally it is much scarcer than the Red Admiral *V. atalanta* but 1993, when there was only a single record, from an observer in Wigan, was exceptional. Equally exceptional was the invasion of 1996; the butterfly arrived in vast numbers in early June, quickly spread throughout Britain and at the peak, from 8.vi to about 20.vi, was the most abundant species in most of the areas recorded. Another wave of abundance followed in August, commencing on 1.viii and gradually lessening through the month; by this time *V. atalanta* was on the increase and for most of the remainder of the season the numbers of the two species were about equal. The concentrations of records on the map into certain areas, and absence from others, are far more an indication of where the recorders happened to be during the 1996 season than a representation of the butterfly's true distribution that year; it would almost certainly have occurred in every square.

At the peak of the migration, observing from a randomly selected open site in Oldham (SD9106) on 9.vi, butterflies were observed passing at a rate of 10 per hour in a north-westerly direction. At another location in Oldham (SD9105) the same day, numbers of them were congregating on a scrap of waste ground bounded by an east-facing wall and nectaring on Oxford ragwort. The species reverted to its normal scarcity in 1997 and was only seen in three 100 m squares in the 7 X 5 km zone, whereas in 1996 it had been seen in 142 such squares.

Broods. Normally immigrants are first seen in June or July, followed by a locally-bred generation in August or September, sometimes continuing in smaller numbers into late September and October. The timings vary greatly from year to year; in the exceptional abundance in 1996 both broods were timed earlier than usual.

Behaviour. Although regularly seen with other large nymphalines, this butterfly does show behavioural differences owing to its centre of geographical range being a warmer clime. The immigrant butterflies regularly pick out the hottest patches of bare soil, sunbaked slopes, stones or unsurfaced roads, and establish territories around them. The species basks more often than the Red Admiral *V. atalanta*, and when feeding generally holds its wings fully open whereas *V. atalanta* more often feeds with wings three-quarters open. Although *V. cardui* will come into gardens to nectar on buddleia *Buddleja davidii*, its predilection for this plant is not as extreme as *V. atalanta*'s and it will use a much greater variety of other nectar sources. During the 1996/7 nectar survey, 36% of *V. cardui*'s recorded plant/days were on buddleia as against 77% of *V. atalanta*'s, and *V. cardui* was noted on 24 other plants, including Guelder rose *Viburnum opulus* on which numbers were observed in the Mersey Valley during the June 1996 immigration. The species does not continue feeding, or flying, as late in the day as *V. atalanta* as its flight temperature threshold is higher. On 11.ix.1994, one example of each species was seen at a small grassland site with some trees at Sinderland Road, Broadheath (SJ7590), in hazy sunshine with clouds and moderate wind: the Painted Lady only flew when flushed, when it made a short flight, landed on the path with wings closed; it was then flushed again, landed on a dead knapweed stem, head downwards, walked down the stem and took up a roosting position on a leaf.

SMALL TORTOISESHELL, *Aglais urticae* (Linnaeus, 1758)

Habitats. This species, in common with the Peacock *Inachis io* and Comma *Polygonia c-album*, has three main requirements: (1) breeding habitat, (2) nectar sources, (3) hibernation sites. These may not coincide, and additionally mating and roosting may occur at different sites (Baker, 1969). Regarding breeding habitat, in spring, following hibernation, the butterflies converge on warm spots where there are nettles. In the Mersey Valley, a favoured habitat is the central north-south drain in Crossford sports field, Sale (SJ7993), where every year in early spring the warm microclimate of the west-facing ridge attracts aggregations of these butterflies and most years hordes of caterpillars follow on the nettles which grow all along the drain. The species will breed in most places where there are nettles (apart from within dense woodland) including agricultural land where nettles often occur in association with farm waste, and around buildings and in otherwise "improved" fields; also waste land, river valleys, etc. In summer and autumn, freshly-emerged butterflies move about in search of nectar and are often seen well away from breeding habitat, including in suburban gardens and parks. Regarding hibernation sites, the only observation during this survey was of numbers of these butterflies and Peacocks *Inachis io*, plus two Herald moths *Scoliopteryx libatrix*, in old air-raid shelters at Bradley Fold, Radcliffe (SD7508) in October 1991. Just to what extent these butterflies rely on man-made structures as hibernation sites is impossible to determine; presumably they must normally select somewhere more natural.

Hostplants (figs.114 & 160). Nettles *Urtica dioica* and *U. urens*.

Broods. There are normally reckoned to be two broods per year as usually butterflies start emerging in late June and fresh examples continue to appear through August and September. The records do not however show very distinct peaks and butterflies are seen continuously throughout the summer; it also seems fairly certain that in some cold years only one brood has been produced; this was noted in the mid to late 1980s. Frequently, in late spring, very small and almost fully-grown larvae are found on the same nettle-bed as a result of a very protracted egg-laying period, so it is probable that sometimes what may appear to be early second-brood butterflies are actually late first (or only) brood. On 12.ix.1995, young larvae were found in the Bollin Valley, Wilmslow (SJ8481). These could have represented an attempted third brood; apparently the exceptional heat of the 1995 summer overrode the butterfly's normal response to photoperiodicity (change in day-length) (Dennis, 1985).

The autumn butterflies hibernate and reappear in March, or April in cold wet springs, and continue flying into May. In 1990 one was seen flying on 18 January in the Mersey Valley, Sale, near the "Priory" woodlands (SJ7992); this was exceptional, and presumably butterflies which are thus awakened too early are capable of going back into hibernation.

Hibernated butterflies are long-lived and sometimes a few overwintered adults can still be seen in June after the early summer brood has started emerging.

Distribution (figs.27, 58, 95 & 141). In good years it is found over virtually the whole area. Although not dependent on migration like the *Vanessa* species, adults of this species and the Peacock are highly mobile and do not form permanent colonies which breed in any one site from year to year; the presence of a breeding population in any particular site in one year does not necessarily mean that there will be one the next. Large fluctuations in numbers occur, probably as a result of weather conditions, and not always as predicted. Following the abundance in autumn 1992 (when "myriads" were observed in suitable localities, notably the Priory, Sale (SJ7992)) many hibernated successfully and there were many records in spring 1993 (including, on 3.v, no fewer than 84 on nettle-beds on the bank of the river Irwell by Littleton Road playing fields, Salford (SD8000, 8001, 8100). The year 1993 was not however a good breeding season and few were seen in the summer and autumn. In consequence, not many were seen after hibernation in 1994, and another poor breeding season, due probably to unsuitable weather in the spring, resulted in the butterfly being almost a rarity in the autumn. Numbers then picked up in 1995, in spite of overwintered butterflies being sparse throughout much of the Manchester area. Taking the Mersey Valley as an example, unusually high numbers of caterpillars were found at sites such as Crossford sports field and the east-west track through Mosley Acre Farm, a short distance away (both SJ7993), followed by an abundance of the butterfly in the summer in a number of Sale sites. For example, during a lunch-time observation on 6.vii, 83 individuals were seen during less than half an hour at the slurry farm on Little Ees Lane, Ashton on Mersey (SJ7793), nectaring on thistles. 1997 was another year of abundance, and in the 7 X 5 km zone this species was the most numerous butterfly that season, a position normally held by the Small White *Pieris rapae*.

Another wave of abundance occurred in 1997; however in this year it was the autumn emergence which was in exceptional numbers. As an example of the abundance of this species that autumn compared to other butterflies, on 10.ix.1997, a warm sunny day, during a 10-minute visit to a patch of waste land approximately 100 m x 70 m in size, in Sale (SJ792916), a former tennis court adjacent to a school playing-field where a flora comprising a number of pioneering plants including many young buddleia bushes had developed, 51 *A.urticae* were recorded as against 3 *Pieris brassicae*, 7 *P.rapae*, 1 *Vanessa atalanta*, 2 *V.cardui* and 1 *P.c-album*. Many of the records in autumn 1997 were of butterflies nectaring on garden buddleias away from breeding habitat; the 3 X 2 km map is much more an indication of the range through which the adult butterflies will wander in search of nutrition than of the breeding range.

Smith (1995) commented on the apparent loss of *A.urticae* from another sports field in Sale (Budworth Road, SJ8091). Investigation during this survey in 1995 proved, however, that the butterfly was in fact breeding that year in that field; it is suggested that the "scarcity" noted by Smith was a result of cyclical fluctuation in abundance caused by climatic variables and/or parasites (Hardy, 1995). There is nothing to suggest that the species has undergone any range reduction.

Behaviour. In the spring following hibernation, the butterflies spend much of the time warming up by dorsal absorbence basking, on the ground, stones and the like, with wings fully open. They are territorial and can be very aggressive; they regularly attack Peacocks *Inachis io* and also queen humble-bees which frequently invade their territories. At this time of year the butterflies use dandelion *Taraxacum officinale* as their main nectar source. When the summer butterflies emerge, they frequently use creeping thistle *Cirsium arvense*; later in the season the flowers of this plant seem to become less attractive, and the butterflies will then travel far in search of nectar and use non-indigenous flowers, chiefly buddleia *Buddleja davidii* and (in some years) Michaelmas daisy *Aster novi-belgii*. On 15.x.1994, during a four-hour walk in ideal weather in the Hazel Grove and Ladybrook Valley areas, three were seen nectaring on a type of single-flowered pinkish-orange chrysanthemum growing in a garden on Ashbourne Road (SJ9285), along with one Red Admiral *Vanessa atalanta*. There was no breeding habitat nearby and the only other butterfly seen anywhere during the walk, either in other gardens or in natural habitat, was another Red Admiral.

Some other nectar sources noted as used by this butterfly, in addition to those recorded in the detailed 1996/7 survey, include hemp agrimony *Eupatorium cannabinum* (Moses Gate (SD7407) in the Croal/Irwell Valley on 27.viii.1995, with the Red Admiral and Small Copper *Lycaena phlaeas*; although this plant occurs quite commonly in the valley it usually seems surprisingly unpopular with butterflies in this area, unlike elsewhere in the country), Himalayan balsam *Impatiens glandulifera* (14.viii.1995) and Lucerne *Medicago sativa* (a number seen near Lowton Common (SJ6397) on 19.viii.1995).

Sometimes these butterflies are recorded in the hills to the north and east, possibly in search of small pockets of nettles amidst the largely grazed grass. On 13.v.1996 a number were seen on hills to the east of Bolton, where there appeared to be very little hostplant/habitat; at grid reference SD758135 five were flying and interacting together and it is wondered whether this behaviour was a form of the "hill-topping" process which some species adopt for mate-location.

CAMBERWELL BEAUTY, *Nymphalis antiopa* (Linnaeus, 1758)

Habitats. All the records are from suburban areas, mostly gardens and one from a school. These are clearly not the butterfly's natural habitats; being well outside the normal breeding range the individuals seen were presumably seeking whatever nutrition was available.

Hostplants. No record of breeding in the area.

Broods. The life-cycle is a single brood overwintering as an adult butterfly. The sightings were of freshly emerged immigrant butterflies, with the presumed exception of the one at Sharples School, which may have been an immigrant from the previos year that survived the winter.

Distribution (fig.28). All but one of the reported sightings were part of the exceptional immigration which reached Britain in 1995. On 2.viii.1995, B.T.Shaw saw two, one at the "Waterside Hotel", Didsbury (SJ8589) and the other in a garden at Lynton Park, Cheadle Hulme (SJ8685). On 20.viii.1995 R.S.Greenwood noted one at Bramhall Close, Milnrow, Rochdale (SD9212). Two further sightings were reported via the Bolton Museum; one was at Barndale Close, Ainsworth (SD7609) on 4.viii.1995 (R.Benham) and the other at Queen's Avenue, Bromley Cross (SD7213) on an unspecified date (D.Lumb). Finally, there is a single report from 1996, which S.P.Garland "identified from a vivid description", at Sharples High School, Bolton, in June, an unlikely date.

Behaviour. Both Shaw's records were of butterflies feeding on fallen fruit. The Didsbury example was utilising rotten pears in the hotel car park: although the fruits were swept up daily the butterfly was reported as remaining at the site three days; the other was on fallen apples.

PEACOCK, *Inachis io* (Linnaeus, 1758)

Habitats. This species has very similar ecological requirements to the Small Tortoiseshell *Aglais urticae*, including the tendency to converge on warm, sheltered spots in the spring after hibernation, such as the centre strip in Crossford sports field in the Mersey Valley (SJ7993) (Hardy 1995). Most open or lightly wooded habitats with nettles can support the species, and like the Small Tortoiseshell it can use patches of nettles growing near farms where "improved" agricultural land renders the habitat unsuitable for most other butterflies. It is more of a woodland butterfly than *A. urticae*; for example, during a walk through Botany Bay Wood, Worsley (SJ7298/7398), an extensive woodland consisting largely of mossland birch interspersed with rhododendron, and some larger trees, and not "managed" for wildlife, on 3.v.1997, in warm cloudy weather, *I. io* was the only butterfly species seen. It occurred in good numbers in spite of there not being any apparent breeding habitat in the dense wood.

In summer the newly-emerged adults require abundant nectar and if this is not available in the breeding habitat they will move elsewhere, including into gardens, to seek it, though not to the same extent as the *Vanessa* species. The only observations of hibernation sites during this survey were in October 1991, when D.Bentley found a number of these butterflies hibernating in old air-raid shelters in Bradley Fold, Radcliffe (SD7508), along with Small Tortoiseshells and two Herald moths *Scoliopteryx libatrix*, and on 3.ix.1995, when two were found in a similar site near Bramhall (SJ8784). The Bradley Fold site was destroyed shortly afterwards. Whilst disused military buildings such as these, together with adjacent regenerated vegetation, certainly do provide valuable overwintering sites for butterflies and other wildlife, it is difficult to make a case for their retention on this account as the butterflies must surely normally utilise more natural sites, which remain undetected.

Hostplants (figs.114 & 160). Nettles *Urtica dioica* and *U. urens*.
Broods. One annual brood, emerging in late July and August. When reared in captivity, whole broods will emerge synchronously; it is likely that this also happens in the wild, as often large numbers of fresh butterflies are seen together, for a short period in habitats with abundant nettles and thistles where they may well have bred. The summer flight period is short and butterflies normally enter hibernation early, presumably as soon as they have acquired enough nutrients to see them through the winter, even though hot weather may continue. Hibernated butterflies appear again in March or April to May, along with the Small Tortoiseshell and Comma *Polygonia c-album*. They are extremely long-lived and very worn individuals are sometimes seen still alive in late June or even July.

Distribution figs.29, 60, 96 & 142). Generally distributed over the area. Normally it is much less abundant than the Small Tortoiseshell but this is not always the case; in 1993 there was an unusually large emergence and the Peacock outnumbered the Small Tortoiseshell in many sites. Not as many reappeared in spring 1994 after hibernation as might have been expected; but the summer emergence was nevertheless good that year, at any rate in the Mersey Valley, where the species again outnumbered *A. urticae*.

Behaviour. In spring the overwintered butterflies need to thermoregulate frequently and are often seen on warm surfaces such as dry paths, stones or pieces of plastic, with their wings fully open and appressed to the substrate, exactly aligned to the sun's azimuth. They are territorial and also frequently interact with *A. urticae*. These interactions are normally aggressive but sometimes attempted courtship of an *I. io* female by a male *A. urticae* has been noted, usually towards the end of the spring flight period; an instance was on 3.v.1992 at the Urmston oxbow (SJ7693) in the Mersey Valley.

Following emergence, the fresh summer butterflies will take nectar from flowers such as thistles if these are available close to the breeding habitat; on the other hand there is evidence that butterflies may disperse far and congregate at nectar sources. On a single buddleia *Buddleja davidii* at Haigh Hall (SD5908) on 6.viii.1995 there were 19 *I. io*; none were seen elsewhere in the vicinity and no breeding habitat was observed nearby; other butterflies present were 5 *A. urticae*, 1 *Vanessa atalanta*, 4 *Pieris rapae* and 1 *P. brassicae*. Even more extreme was the aggregation in a thistle-field on the Pennines just east of Dovestone reservoir (SE0202) on 2.viii.1995 when the distribution was 91 *I. io* with 14 *A. urticae*, 1 *V. atalanta*, 3 *V. cardui*, 3 *P. rapae* and 1 *Maniola jurtina*. Here it was possible that the *I. io* had bred nearby but more likely that their behaviour was a modification of "hill-topping" for the purpose of obtaining a pre-hibernation nectar source, and that they had come in from some distance; several individuals were seen flying over totally unsuitable terrain higher up the hill presumably in search of nectar.

On 9.viii.1997, a hot sunny day, during an afternoon of observations covering suburban and rural sites in approximately equal proportions, in the vicinity of Blackrod (SD6010, 6011, 6110, 6111 and 6210), of a total of 102 *I.io* observed taking nectar, 38 were using buddleias, mainly in gardens, 54 were using creeping thistle, in open country, and only 10 were using other sources (privet 4, spear thistle 3 and teasel 3).

COMMA, *Polygonia c-album* (Linnaeus, 1758)

Habitats. The most favoured habitats are woodland edges and rides, and open patches within woods, including in the river valleys, even quite near the city centre. Suburban green spaces, including parks with some trees, are sometimes suitable; Crowcroft Park in Levenshulme (SJ8795) is an example. The late summer butterflies are often seen away from breeding habitat in search of nectar.

Hostplants (figs.114 & 160). Presumably nettle. Elm *Ulmus* spp. is a potential alternative but in view of its present scarcity is unlikely to be significant.

Broods. Hibernated butterflies appear in March or April and normally continue to fly until early May. The last of the overwintered butterflies are sometimes very long-lived and may even overlap with the mid-summer emergence of the paler, less angular form *hutchinsoni*; on 12.vii.1995 both a very worn butterfly and a fresh *hutchinsoni* were seen in Sale. The mid-summer *hutchinsoni* flight is however very short-lived, occurring in late June and July. Non-*hutchinsoni* butterflies emerge from the second half of July (e.g.17.vii.1995, Sale) and there is a later emergence stretching from August to October. Of the three hibernating nymphalids, *P. c-album* usually is the one to continue to be seen flying latest in the autumn before hibernation.

Distribution (figs.30, 61, 97 & 143). Although there had been earlier records from the Bollin valley, prior to 1987 the species was very scarce or absent in most of Greater Manchester. It expanded its range in 1987-1992, reaching most of the area, including the city centre, but with a clear southern bias. A peak in abundance seems to have occurred in 1990, large numbers of overwintered butterflies being recorded in March that year in Heald Green (SJ8485)

(B.T.Shaw, pers.comm.) and the Mersey Valley, followed by a slight reduction in the following two years. Then, in 1993, numbers in the spring appeared surprisingly low in relation to the previous autumn, implying a poor winter survival rate of the hibernating butterflies. Numbers did not pick up during the year and there were very few in spring 1994. There was however a good autumn emergence that year and in 1995 numbers appeared to be back to normal. The scarcity in 1992-3 could have been a cyclical fluctuation; I wonder however whether the spread of this species and subsequent fluctuations in abundance might be correlated with garden-escape Michaelmas daisies *Aster novi-belgii*. In some districts, notably the Mersey Valley and parts of the Croal/Irwell Valley, where there are large swathes of this plant; several nymphalines, especially this species and the Small Tortoiseshell *Aglais urticae*, frequently use it as a nectar source in the autumn and certainly *P. c-album* often seems more numerous in areas where this nectar source is available. Few indigenous flowers provide nectar for butterflies as late as October, and it does appear that the late emerging *P. c-album* rely heavily on this plant. In 1992 the Michaelmas daisies flowered rather early but the main emergence of *P. c-album* did not occur until October by which time the flowers were nearly over and possibly many butterflies could not obtain enough nutrition to see them through the winter.

Behaviour. Following hibernation the butterflies frequently bask on warm ground with wings fully open. They are often fiercely territorial, and sometimes aggressive to other species as well as their own. At Partington (SJ7291) on 21.vi.1995, a worn one was observed to attack two Large Skippers *Ochlodes venata* which invaded its territory, but eventually they drove it out. The early summer *hutchinsoni* butterflies tend to remain close to the breeding habitat and often settle on trees, sometimes sharing habitat with the Speckled Wood *Pararge aegeria*. The later summer butterflies will move out of the habitat to seek nectar sources including buddleias *Buddleja davidii*, for which they will come into gardens and parks, though not to the extent that some other nymphalines do, especially the Red Admiral *Vanessa atalanta*. Privet *Ligustrum ovalifolium* is sometimes used. The species will also take nectar from indigenous flowers when available, mainly composites such as thistles *Cirsium* spp, and in spring dandelions *Taraxacum officinale*. Fleabane *Pulicaria dysenterica* (not a common plant in the Manchester area) was being used on the south edge of Ringley wood (SD7605) in the Croal/Irwell Valley on 27.viii.1995; this is a first-class woodland-edge habitat and Red Admirals were also using it.

The use of Michaelmas daisies by the autumn butterflies has been discussed above. Elsewhere they have been reported using ivy blossom (Wilmslow, 1995) or fallen fruit such as apples and pears (R.L.H.Dennis, pers. comm.). In the Mersey Valley several were seen in 1991 and 1992 feeding on ripe blackberries on one particular bramble bush near the Stretford sewage farm (SJ7893); this behaviour occurred again in 1997 but no further instances of these butterflies obtaining nutrients from a source other than flower nectar were noted during the survey.

DARK GREEN FRITILLARY, *Argynnis aglaja* (Linnaeus, 1758)

Habitats. Of the three reported sightings, two were on revegetated former tips and the third in a wooded site.

Hostplants. Violets *Viola* spp. There is no evidence of breeding in the mapped area.

Broods. Two of the sightings were in July (no date was given for the third). This tallies with the species's normal life-cycle of a single brood in July/August.

Distribution (fig.31). Sightings have been reported by K.McCabe from the Flixton fly-ash tip (SJ7393) (1993) and by a member of the Manchester Field Club from Chorlton tip (SJ8192) (1992). Both were experienced observers who were confident that the identifications were correct. The third is a report passed to the Manchester Wildlife Trust by a recorder from Alkrington Wood (SD8604) as *Boloria selene*; however a photograph produced in support of the claimed sighting was of *A. aglaja*. There are no records of this species from the area in the historic data, though Blackie (1946) mentions a possible sighting in Wythenshawe (see page 11).

Behaviour. The species is a strong flier and, although it is not normally regarded as migratory, the sightings must presumably have been strays from a considerable distance away: the nearest known breeding colonies are in the Peak district (Cunningdale, Lathkill Dale) and the Lancashire coast south of Southport.

Family NYMPHALIDAE (sub-family SATYRINAE)

SPECKLED WOOD, *Pararge aegeria* (Linnaeus, 1758)

Habitats. Well known as a shade-loving butterfly, the prime requirements of this recent colonist are habitat with at least semi-mature trees giving shelter plus a mixture of sunlight and shade. The mature woodlands in the Bollin Valley (SJ7884, 7983 and others outside the boundary) have provided suitable habitat; so have many of the more recently planted woodlands further north including in the Mersey Valley. Other new habitats have developed from abandoned railways, for instance the former Walkden to Leigh line (SD6901 to 7201), which now forms the equivalent of a linear east-west strip of fairly sheltered woodland ride several miles long, as trees have grown up on either side for most of its length. On 27.viii.1995, in seemingly less than ideal weather (windy with heavy showers alternating with periods of sunshine) during a walk along this track, the present species and the Red Admiral *Vanessa atalanta* were both present in exceptional numbers, an instance of a particular combination of habitat and climatic variables fortuitously benefiting two not closely related butterfly species with very different biology, one being a recently established resident with specific requirements and the other an opportunist immigrant.

In 1995, 1996 and 1997 *P. aegeria* was found in two suburban parks in Sale (Worthington Park (SJ7991/2) and Walton Road Park (SJ7890)) where the formally planted trees and ornamental shrubberies appear to be sufficiently like a woodland edge to satisfy it, even though the grass is regularly mown very short. It also appears to have established itself in some of the large suburban gardens in the west of Sale (SJ7791) where there is sufficient tree canopy to provide the required shade. On the other hand, observations in 1995 suggested that not every apparently suitable habitat within its range supported the species; on 9.vii.1995 it was found in what had seemed an unlikely wood in Brooklands Estate (SJ7989), isolated by playing fields and high-density housing and a mile from the nearest previously-known habitat, but searching more likely-looking woods further to the south, along the Fairywell Brook (SJ8088) the same day failed to find it.

Hostplants (figs.116 & 162). Grasses, which must be growing in at least partial shade. The only egg-laying observation during the survey was of a female laying on a small scrap of grass (species undetermined) on a mound of soil by the old lane adjacent to the Priory nature reserve, Sale (SJ8092). If in fact the butterfly is breeding in the parks in Sale, as it appears to be, the caterpillars must be able to survive severe mowing. The hostplant/habitat maps plot areas of woodland, as tree cover appears to be the important limiting factor rather than the kind of grass and its quantity. See Shreeve (1984).

Broods. The voltinism is very confusing (Shreeve, 1986). There appears to be a very long, staggered appearance of each brood due to the species overwintering in two different stages of development. The first butterflies are usually on the wing in the first week in May, though in 1997 they were seen as early as 9.iv. The April and May butterflies appear to represent the first part of the spring brood (from overwintered pupae); there is then a gap with no sightings followed by the second part of this brood (from overwintered larvae) which is out in June.

There are further emergences (first and second parts of the second brood?) in July and August/September; some butterflies later in September and in October possibly represent a third brood.

Distribution (figs.32, 66, 98 & 144). The species was virtually unknown in the area until 1990 and since then has undergone a spectacular colonisation from the south and west (Hardy, Hind & Dennis, 1993). The Bollin Valley, to the south, was colonised early, with several records in 1990 and numerous sightings in most of the woods in 1991. Further north in the Mersey Valley, the first sightings were also in 1990 (the Priory, Sale (SJ7992) south of the river, on 3.viii, and the Chorlton Ees (SJ8093) north of the river, on 8.viii), but colonisation was slower. A viable population established fairly quickly in the Priory, but the initial sighting in the Chorlton Ees was not followed by any more for several years, and it was probably 1995 before the butterfly was firmly established there. Similarly, it was first seen in the lane adjacent to the Stretford sewage farm (SJ7893) in 1992 but there were no further sightings there until 1994. These records indicate that of these three Mersey Valley sites one was a quick colonisation and the other two not immediate, the first sightings being strays - wandering butterflies exploring new habitats but not yet being able to found a colony. Alternatively, the breeding populations may have been established but in too low numbers to be apparent in the intervening years. Findlay & Tilley (1995) refer to *P. aegeria* as a "rather sedentary butterfly". This has certainly not been the case in Greater Manchester during this survey; to colonise as it has done the species must have very considerable powers of dispersal, and indeed individuals on passage have frequently been seen moving through unsuitable habitat, and crossing motorways (for example, the M.63 in Sale (SJ7993) on 26.ix.1995).

By 1995 the butterfly had spread to the Worsley area (SD7400/7501), west of Manchester, in good numbers, and had reached the Bolton and Wigan areas though in much lower numbers. On the other hand, there have been surprisingly few records from the Stockport vicinity; in the north and east of the mapping area many apparently suitable sites have shown no sign of the butterfly. This situation could well change if the range extension continues; although climate change has been suggested as a cause of this range extension, the indications from the present survey are of a steady increase from year to year since 1990 in spite of fluctuations in the weather. Some exceptional summers have probably seen the establishment of new large source populations from which other large and small habitats are continually being colonised. The smaller ones will be "sinks" - temporarily colonised (Dennis, pers. comm.)

Behaviour. Ideal flying conditions are a mixture of sun and shade. As an illustration, on 13.viii.1995, in Nan Nook Wood in Wythenshawe Park (SJ8090), following an overcast morning in which very few butterflies had been flying, the sun broke through at mid-day and almost immediately seventeen *P. aegeria* appeared in a normally damp area in the centre of the wood (which however had dried out in the unusual heat of 1995), where gaps in the tree canopy admitted some light. The only other butterflies present were two Green-Veined Whites *Pieris napi*. This site is a unique habitat consisting of unmanaged mature woodland and swamp bounded to the south by a large expanse of formal parkland and to the north by housing.

Although its optimum conditions for flight are as above, the species does sometimes fly in overcast weather, and was even noted flying in slight rain during a period of exceptional abundance at the end of August 1995. It has also been seen flying in dense closed-canopy woodland (albeit on a warm, sunny day) on the M.63 motorway embankment close to the Mersey Valley visitor centre (SJ8092) on 6.v.1995.

It regularly thermoregulates by basking with wings open, in a small patch of sunlight, usually on a leaf above ground level, and is very territorial. On 23.viii.1994, at 6 p.m., one was seen basking on a piece of white paper on the ground at the edge of Bradley Lane, Stretford, below one of the low lime trees which line this road.

The butterfly is normally seen basking or flying and very rarely takes nectar. The few observations of nectaring during the survey suggest that Michaelmas daisy *Aster novi-belgii* is probably the most favoured source.

WALL, *Lasiommata megera* (Linnaeus, 1767)

Habitats. This butterfly favours warm, dry sites, such as waste ground, and sloping valley sides especially with a south-facing aspect. Abandoned railway sidings are popular. The main differences between the habitat requirements of this species and the Meadow Brown *Maniola jurtina* are that *L. megera* prefers areas of patchy grass amidst bare soil, including verges and grass next to hedges, and will not use thick tall grass; however the requirements of the two species do overlap and it is rarely possible to state categorically that any particular grassland site could support the one and not the other. Also, although the present species uses a very different hostplant, its habitats are remarkably similar to the Small Copper's *Lycaena phlaeas*, and, interestingly, both species are orange/brown in colour. Observations on the Mersey bank (SJ7892/3) show that *L.megera* can tolerate a certain amount of mowing; the bank is normally mown at least once per annum.

Hostplants (figs.102 & 148). Grasses; the precise species used in the Manchester area have not been identified. Females prefer to lay on small clumps of fairly short grass in warm, sheltered spots.

Broods. There are always two broods per year; the second is always more numerous than the first and of longer duration. The first brood appears in May, usually from the middle to the end of the month, but in 1990, a year with a warm spring, numbers of butterflies were seen on 1.v and the females had already started laying. The second brood is normally out in late July and August, though in some cold summers, for example 1985, 1986 and 1987, it has not started until mid-August and has continued into September and in 1986 early October. In some years, notably 1989 which had a warm summer and in which the species was exceptionally abundant, a few fresh butterflies in late September and early October appear to have been a partial third brood.

Distribution (figs.33, 67, 99 & 145). The map indicates a general distribution over most of the area apart from the hilly ground in the north and east. Unfortunately it appears that the climatic or other factors which favour the previous species, *P.aegeria*, affect this one adversely and during the last few years of the survey it appeared to undergo a considerable reduction in numbers, this being especially noticeable in 1997. It is however too early yet to be certain whether this is a range contraction or part of a cyclical fluctuation in abundance.

Frequently in years when the Small Copper *Lycaena phlaeas* has been abundant, such as 1989, *L. megera* has also been, though in 1994, despite the timing of the second brood being synchronised with that of *L. phlaeas*, *L. megera* was much more abundant. In 1995 and 1996, very poor weather in late spring resulted in low numbers, even though 1995's spring was followed by a hot, dry summer; the spring brood was in low numbers and was over very quickly, and the summer brood was of no more than average abundance.

Behaviour. This butterfly is well known for its habit of basking on warm surfaces; however only very rarely have I observed one to bask on a wall, as the English name suggests! Much more popular, in my experience, are bare earth and stones, including railway ballast. During the period of exceptional abundance in 1989, on 5.viii many of these butterflies were seen late in the afternoon basking on the iron rails of the little-used mineral railway (since completely abandoned) in the south-west of Carrington Moss (SJ7391).

Interactions between individuals of this species, and with other butterflies, are often aggressive, but not always. At times during this survey two males have been seen to select adjacent basking positions, and likewise on 14.viii.1995 one was observed to select a position, apparently deliberately, immediately beside a basking Small Tortoiseshell *Aglais urticae* on the Mersey bank (SJ7892).

GATEKEEPER, *Pyronia tithonus* (Linnaeus, 1771)

Habitats. Favoured habitats are hedgerows, field edges, woodland edges and scrub, usually with abundant brambles which are a popular nectar source, and often with tall grass and bracken. Some shelter-belt is essential. Around Wigan, biotopes associated with former mines and reclaimed tips, with warm sloping sides, form ideal sites. The butterfly also occurs in numbers in a deep, disused railway cutting, now a footpath, at "Red Rocks" to the north of Wigan (SD5807/8), and in much lesser numbers along the side of the Leeds to Liverpool canal (SD5907/8). In some sites, notably one near the railway and the Manchester ship canal near Flixton (SJ7293), habitat stratification with the Speckled Wood *Pararge aegeria* has been observed; the two species occur in very close proximity on the river-bank on the south side of the railway embankment and a lower-level wooded track on the north side. The shadier habitat suitable for *P. aegeria* is at a slightly lower level, and although the two species can be seen within a few yards of each other they do not normally overlap.

Hostplants (figs.102 & 148). Grasses. The precise species used in the Manchester area have not been identified.

Broods. One annual brood, flying from mid-July into August. The flight period is shorter and later than the Meadow Brown's *Maniola jurtina.*

Distribution (figs.34, 69, 100 & 146). *P. tithonus* is localised, and has generally been a rare species, apart from in the west, where there are established and increasing colonies around Wigan, Leigh and adjoining districts. It is however clearly spreading eastwards and in 1994 and 1995 small colonies were discovered in Carrington Moss (SJ7391) and Partington (SJ7291/2). At the latter site on 23.vii.1994, only 4 *P. tithonus* were seen as against 122 *M. jurtina*, whereas in its longer-established sites north of Wigan, particularly in the Haigh area (SD5808, 5908) *P. tithonus* sometimes outnumbers *M. jurtina*. Single individuals have been noted in unexpected sites where the species had never been seen previously, such as on the Mersey bank, Stretford (SJ7993) on 16.vii.1992, on the edge of a horse-field in Carrington (SJ7392) on 23.vii.1994, and on a scrap of waste land in the centre of Sale (the former King Street, SJ7892) on 31.vii.1994. Evidence from K.McCabe suggests that the species may already have been established at the western end of the Mersey Valley at Flixton (SJ7393) by 1992; it has since appeared in other parts of the Valley, including near the Stretford sewage farm (SJ7893) in 1996, and by the Sinderland Brook (SJ7590) further south and just outside the 7 X 5 km area, in 1995. There have been reports from other observers (B.T.Shaw, G.Crossley, J.Thompson) from Heald Green (SJ8485), Bramhall (SJ9086) and near the airport (SJ8083); a sighting was also reported in Hyde (SJ9594) in 1994, and in 1995 a colony was found near the Ashton under Lyne sewage farm (SJ9297). The evidence suggests a recent extension in range, analogous to those of *Thymelicus sylvestris* and *Pararge aegeria*, but from a different direction (from the west rather than from the south) and less rapid (Hardy, Hind & Dennis, 1993).

Behaviour. Bramble *Rubus fruticosus* agg. is a popular nectar source although the few observations in the 1996/7 nectar survey do not show it. This butterfly is more lively than *M. jurtina* and more inclined than that species to bask with wings open.

MEADOW BROWN, *Maniola jurtina* (Linnaeus, 1758)

Habitats. A very tolerant species occurring in most places with a reasonable amount of not too coarse grass. It does like some shelter, and thus frequents field edges, gully sides, and grass areas in scrubland. Woodland edges are favoured and frequently the butterflies roost in the adjacent trees. In general, high ground and warm, dry slopes are less suitable than damper sites; the species does however occur in a small hollow at the top of Holcombe Hill, at an altitude of 350 m, north of Bury (SD7716). The butterfly prefers medium-length grass; it can tolerate a certain amount of mowing such as the annual cut in hay-meadows or on the river-banks in the Mersey Valley, but not mowing to the extent practised in sports fields and individuals occasionally seen flying over such sites are presumably on passage. Microdistribution is by no means even; in the most favoured spots it can occur at a very high density, but it will also be present in ones and twos in surrounding less suitable habitat, presenting a good case for the use of small-scale mapping.

M. jurtina's habitat tolerance range is very similar to that of the Large Skipper *Ochlodes venata* and very frequently the two species are found together. It also sometimes occurs with the Small Skipper *Thymelicus sylvestris*, though the latter tends to prefer hotter, drier sites: for instance, on the hot, dry, sparsely vegetated slope beside the A.538 road near the airport tunnel (SJ8083), *M. jurtina* exists in very small numbers compared to *T. sylvestris*.

Hostplants (figs.102 & 148). Grasses. The precise species used in the Manchester area have not been identified.

Broods. It is generally accepted now that this butterfly is always single-brooded with a staggered emergence giving a very long flight period, beginning in mid to late June and continuing through July and August, with sometimes a few individuals surviving into September.

Distribution (figs.35, 70, 101 & 147). Most years during its flight period this is the commonest butterfly and virtually guaranteed to be seen in any suitable spot. Although it seems fairly sedentary it clearly must move a fair amount to be able to colonise sites as readily as it does. (Shreeve, Dennis & Williams, 1996). Occasionally butterflies are seen well out of habitat; instances during this survey include one on a busy main road in a built-up area near Leigh (SJ6699) on 16.vii.1995, and two instances of disoriented butterflies entering a marquee erected in the middle of a school playing field in Flixton (SJ7394), in connection with Mersey Valley summer activities, on 31.vii and 1.viii.1995. Transient sites near the city centre have been readily colonised, including the abandoned Ardwick railway yard (SJ8697), patches of waste land near Victoria station (SJ8499), and the strip of waste ground beside New Allen Street, Ancoats (SJ8599) created by the demolition of the former raised railway into the Oldham road goods station. Former tips are also colonised. The Stretford tip in the Mersey Valley (SJ7893/4), following landscaping and covering with topsoil containing much grass seed, has developed into a wide, open, undulating grassland with no trees or shrubs except on the northern edge, where it is bounded by a strip of semi-mature woodland, and just a few trees on the southern side. In the first years of its settling, the site held few butterflies; in 1993 and 1994 *M. jurtina* was numerous only in the shelter-belts provided by the trees; in 1995 and 1996 hordes were to be seen all over the main grassland.

In 1997 records of this species in the 7 x 5 km zone were much fewer than normal. This could have been a result of more of that year's recording being done in suburban rather than rural squares, or it could have represented an unexplained reduction in the butterfly's numbers that year, or both. It does appear that the other grass-feeding species, *T.sylvestris, O.venata, L.megera* and *P.tithonus*, were similarly affected.

Behaviour. Normally the species can exist at high densities and numbers can be seen nectaring together, frequently using thistles *Cirsium* spp., brambles *Rubus fruticosus* agg. or tufted vetch *Vicia cracca*. It often nectars in company with the Large and Small Skippers, but otherwise its behaviour is very different from theirs: *M. jurtina* tends to skulk in the long grass and take short, slow flights, whereas *O. venata* prefers to perch on adjacent taller vegetation. Sometimes however *M. jurtina* can be very territorial and aggressive; it will attack larger butterflies including large nymphalines and pierids. Though it is not usually aggressive when feeding, and not usually a frequenter of buddleias *Buddleja davidii*, one was observed in a "violent" interaction with a Peacock *Inachis io* on a buddleia near "Kirkwood Cottage", Chadkirk (SJ9390) on 29.vii.1995. On 16.vii.1995, one was seen to chase a bee. Other instances of atypical behaviour noted during this survey include one seen feeding on horse dung in the Tame Valley, Denton (SJ9395); the same day one was seen nectaring on privet. On 2.viii.1995, a small patch of cultivated marjoram *Origanum vulgare* in the garden of the "1937 social club" near Greenfield (SE0003) was observed to have drawn this species well out of its usual habitat.

This species is more tolerant of weather conditions than any other butterfly and will fly in almost any conditions including quite heavy rain.

SMALL HEATH, *Coenonympha pamphilus* (Linnaeus, 1758)

Habitats. This species is very localised, though it can be numerous when found, and some of its moorland habitats are suitable for no other butterfly. Favoured haunts are sheltered pockets and warm slopes or gullies; most of the open moorland has been severely degraded by a long history of sheep-grazing and is too impoverished to support any butterflies. A number of known sites are near reservoirs, the east bank of Greenbooth reservoir (SD8517), near Norden, being a good example.

Hostplants. Grasses. The precise species used in the Manchester area have not been identified.

Broods. In the literature (see, for example, Emmet & Heath (1989) and Pollard & Greatorex-Davies (1997), this butterfly is mentioned as usually single-brooded in the north of the country, and multi-brooded elsewhere. My own observations while making the survey suggested that in the Manchester area it was single-brooded, as all but one of my sightings have been in June. The exception, a worn individual seen in the Roch Valley (SD9013), near Rochdale town centre, on 30.vii.1995, seemed more likely to be a late straggler from a single extended brood than a second brood butterfly, as I saw no others that day although I searched much likely habitat and the weather was suitable. Records from other observers, mostly in the Oldham area, suggest however that either the single brood is very extended, as in the case of the Meadow Brown *Maniola jurtina*, or there is more than one brood, as sightings have been reported from May to September.

Distribution (figs.36 & 71). Colonies occur in the north and east of the area, on the Pennine fringe. Unlike most Manchester butterflies this species appears to be on the decline, retreating eastwards and becoming more restricted to high moorland. B.T.Shaw (pers.comm.) mentions former colonies in the Heald Green area (SJ8485), which appear to have flourished in the 1980s but died out in 1991; and the species was reported in 1991 and 1993 from the embankment beside the A.538 road near the airport tunnel (SJ8083); however I searched this site thoroughly in 1994 and 1995 and was unable to verify its continued occurrence; the same applies to the various reports in the north-west of the mapped area. In 1997, however, there were reports from two observers of single sightings on the reclaimed Adswood tip, Stockport (SJ8887), a large expanse of open grassland; the butterfly therefore may still be hanging on at very low density in parts of the south and south-east.

There have been two difficulties with recording this species: (1) My own observations are not as comprehensive as with some other species as the weather during the flight period was frequently not ideal and working the hilly terrain was difficult and slow; I was unable to record the species at all in 1994, 1996 and 1997; (2) a number of vague reports have been received from casual observers, which on investigation have turned out to be misidentifications; although I have aimed to exclude doubtful records it is possible that some incorrect ones may have slipped through.

Behaviour. The butterflies are usually seen flying short distances low over the grass, and rarely take nectar. On 18.vi.1994, however, a number were nectaring on tormentil *Potentilla erecta* and a white bedstraw, probably *Galium saxatile*, at Greenbooth reservoir in the late afternoon sunshine. Although normally found in isolated colonies, vagrants can travel considerable distances, as evidenced by the one seen at Rochdale on 30.vii.1995 or by R.L.H.Dennis's finding one in the middle of a barley field at Alderley Edge (Cheshire, outside the recording area) in the early 1980s.

Disused canal – Manchester, Bolton & Bury Canal, Nob End, SD7506, 31 vii. 1983. The canal provides a "corridor" for dispersal; habitat for a number of species including *L.phlaeas* and *A.urticae.*

Open moorland above Norden reservoirs, SD8515, 13. x. 1985. Virtually all tree cover has been lost and the biotope is severely degraded by sheep-grazing, but some *P.napi* and *C.pamphilus* habitat remains.

Open countryside near Woodford, SJ8981, 26. xii. 1983. "Cheshire Plain" – type agricultural land; the hedgerow and tree-line provide a potential flyway but the large "improved" fields are not suitable butterfly habitat.

A "wild-flower" meadow in a city park, Fletcher Moss Park, Didsbury, SJ8490, 13. iv. 1985. Habitat for Hesperiids and Satyrines.

The Mersey Bank, Stretford, SJ7892/3, undergoing reconstruction in July 1987. Much high-quality habitat was lost in this process, but has since recovered. See page 13.

The Mersey Bank, Stretford, SJ7892/3, slightly to the west of the location of the previous photograph, in May 1997. 18 species have been recorded along this stretch during the survey.

The reclaimed Stretford tip, SJ7893/4, iii. 1997. A fine example of habitat re-creation, quickly colonised by *T.sylvestris, O.venata, L.phlaeas, P.icarus, A.urticae, I.io and M.jurtina;* further species occur in the shelter-belts bordering the grassland.

Mature hollies in an established suburban garden on Dane Road, Sale, SJ7992, iii. 1997. Breeding habitat for *C.argiolus.*

Formally planted mature trees and shrubs in Southern Cemetery, SJ8292, iii. 1997. The grass in such a site is regularly mown short and does not form ideal habitat; however *P.aegeria* and *A.cardamines* breed, as do *L.phlaeas* and *L.megera* in a more open section nearby. The habitat is potentially suitable for *C.argiolus.*

Manchester Cathedral, SJ8398, iii. 1997. No breeding habitat, but *P.rapae, V.atalanta* and *A.urticae* have been recorded nectaring on a buddleia bush in the centre foreground.

Disused railway at New Allen Street, Ancoats, SJ8599, iii. 1997. A regenerating site, probably short-lived, but providing habitat for *P.brassicae, P.rapae, P.icarus, V.atalanta, V.cardui, A.urticae, I.io, L.megera, M.jurtina.*

Worsley Moss, SJ7298, iv. 1997. A former rich habitat with *C.tullia* and *C.rubi* recorded last century; the area to the left of the ditch has been completely degraded by commercial peat extraction, to the right is regenerating birch scrub. *I.io* recorded.

An active railway line in Urmston, SJ7694, iv. 1997, forming a "corridor" for species dispersal through a densely built-up area.

Experimental tree-planting by the local authority at the Flixton fly-ash tip, SJ7393. 21 species have been recorded at this site.

Waste land and demolition at Verdon Street and "Scotland", behind Manchester Victoria station, SJ8499, iii. 1997. Pioneering vegetation and buddleias provide hostplant-habitat and nectar sources for several butterfly species.

Former tree nurseries on the Sale-Northern Moor boundary, SJ8090, vi. 1997, being destroyed for a housing development. 12 species formerly recorded; see pages 11 and 13.

BURNET MOTHS - family ZYGAENIDAE

As the day-flying Burnet moths are a very conspicuous feature of Manchester's insect fauna and are regularly seen during butterfly-recording walks, often in the same habitats as several butterfly species, I include accounts here of the two species recorded in the area. They are not however included in any of the statistics in this work.

NARROW-BORDERED FIVE-SPOT BURNET, *Zygaena lonicerae* (Scheven, 1777)

Habitats. This moth is likely to occur in any rough grassland, waste land or regenerating site where there is a good growth of its leguminous hostplants and nectar flowers, especially thistles *Cirsium* spp. It often co-exists with the Large and Small Skippers *Ochlodes venata* and *Thymelicus sylvestris* and the Meadow Brown *Maniola jurtina*.

Hostplants. Legumes, particularly meadow vetchling *Lathyrus pratensis*, clovers *Trifolium pratense* and *T. repens*; in some sites it appears to use tufted vetch *Vicia cracca*.

Distribution (fig.169). The moth is found in suitable sites throughout most of the mapped area, including close to the city centre, but excluding the hilly ground in the north and east. It regularly abounds in myriads when its habitats are at their most suitable. Where habitat succession is unchecked, it can be shaded out. This happened on the south-west facing slope of the motorway embankment to the north of Crossford sports field, Sale (SJ7993), where it occurred in vast numbers in the mid-1980s but completely died out as the adjacent woodland matured and scrub on the bank reduced the extent of available habitat. It is however very quick to colonise "new" habitats and its range appears to have extended.

Behaviour. The slow, buzzing flight and bright colouration make the Burnets very easy to see even in dull weather. They regularly take nectar from flowers, often several of them on the same plant, or rest fully exposed on low-growing vegetation.

SIX-SPOT BURNET, *Zygaena filipendulae* (Linnaeus, 1758)

Habitats. Rough grassy sites, waste places, with some regenerating scrub. In each of the few sites where it has been found, *Z. lonicerae* also occurs.

Hostplants. Legumes, particularly bird's-foot trefoil *Lotus corniculatus* (Horwich), possibly tufted vetch *Vicia cracca* (Ardwick) and clovers *Trifolium* spp.

Broods. One annual brood, in July. Normally this species emerges slightly later than *Z. lonicerae*; there have not been sufficient observations within the Manchester area to confirm whether or not this is true here.

Distribution (fig.170). Until 1994 I was convinced that the only Burnet in the Manchester area was *Z. lonicerae*. That year, however, I came across a small colony of *Z. filipendulae* co-occurring with *Z. lonicerae* at a site adjacent to Red Moss, Horwich (SD6310), where the former ballast in an abandoned railway yard had attracted limestone flora including bird's-foot trefoil. The site also supported a strong colony of the Common Blue *Polyommatus icarus*. Shortly afterwards, *Z. filipendulae* was found at the Shell pond, Carrington (SJ7492); then the following year on waste land by Temperance Street, Ardwick (SJ8597), less than a mile south-east of Manchester city centre, again along with *Z. lonicerae* but in much lower numbers than that species. Further sites were found in 1996 including near Flixton (SJ7393) at the west end of the Mersey Valley, where again there has been an introduction of limestone flora, through capping a former fly-ash tip with subsoil from the Peak District. It is possible that the moth is a recent colonist to the Manchester area; alternatively it may have been overlooked in the past or passed over as *Z. lonicerae*.

Behaviour. Similar to *Z. lonicerae* and the two species have been found nectaring together on the same plant of tufted vetch at the Ardwick site.

Table 1. RECORDS OF BUTTERFLY SPECIES IN THE 7 X 5 KM ZONE IN 1994, 1995, 1996 AND 1997.

NUMBERS OF 100 M SQUARES IN WHICH EACH SPECIES WAS RECORDED.

SPECIES	1994	1995	1996	1997
Thymelicus sylvestris	41	62	79	24
Ochlodes venata	85	102	14	17
Gonepteryx rhamni	8	1	7	9
Pieris brassicae	73	84	49	123
Pieris rapae	169	270	230	241
Pieris napi	61	112	100	108
Anthocharis cardamines	57	69	111	69
Lycaena phlaeas	31	30	24	22
Polyommatus icarus	14	13	24	15
Celastrina argiolus	3	1	4	13
Vanessa atalanta	21	32	49	26
Vanessa cardui	4	3	142	3
Aglais urticae	37	173	209	305
Inachis io	44	45	140	179
Polygonia c-album	26	32	31	47
Pararge aegeria	22	67	91	92
Lasiommata megera	39	27	26	10
Pyronia tithonus	1	0	4	1
Maniola jurtina	193	203	141	49

Table 2. NECTAR SOURCES NOTED DURING 1996 AND 1997

PLANT	SPECIES	SPP/DAYS
Creeping Thistle *Cirsium arvense*	17	175
Buddleia *Buddleja davidii*	14	282
Ragwort *Senecio jacobaea*	14	98
Michaelmas Daisy *Aster novi-belgii*	14	64
Bramble *Rubus fruticosus* agg.	13	52
[+Bramble (fruit)	1	1]
Red Clover *Trifolium pratense*	10	31
Black Knapweed *Centaurea nigra* agg.	10	27
Dandelion *Taraxacum officinale* agg.	8	89
Garden Privet *Ligustrum ovalifolium*	8	18
Buttercup *Ranunculus acris/repens*	8	13
Hawkweed/Hawkbit/Catsear *Hieracium* sp.etc.	8	9
Sweet Rocket *Hesperis matronalis*	7	16
White Clover *Trifolium repens*	7	18
Tufted Vetch *Vicia cracca*	5	22
Ice-plant *Sedum spectabile*	5	17
wild Brassica *Brassica* sp.	5	15
Spear Thistle *Cirsium vulgare*	5	14
Oxford Ragwort *Senecio squalidus*	5	10
Himalayan Balsam *Impatiens glandulifera*	5	8
Fool's Parsley *Aethusa cynapium*	5	8
Lavender *Lavandula officinalis*	4	16
Cuckoo Flower *Cardamine pratensis*	4	10
Coltsfoot *Tussilago farfara*	4	6
Lucerne *Medicago sativa*	4	5
Ox-eye Daisy *Leucanthemum vulgare*	4	4
French Marigold *Tagetes patula*	3	7
Garlic Mustard *Alliaria petiolata*	3	6
Hedge Bindweed *Calystegia sepium*	3	6
Great Willowherb *Epilobium hirsutum*	3	6
Hogweed *Heracleum sphondylium*	3	4
(unidentified yellow Crucifers)	3	4
Rosebay Willowherb *Epilobium angustifolium*	3	4
Daisy *Bellis perennis*	3	4
Red Valerian *Centranthus ruber*	3	4
Yarrow *Achillea millefolium*	3	3
Heather *Calluna vulgaris*	3	3
Comfrey *Symphytum officinale*	3	3
Hawthorn *Crataegus monogyna*	3	3
Honesty *Lunaria annua*	3	3
Bluebell *Endymion non-scriptus*	3	3
Black Horehound *Ballota nigra* (?)	3	3
Bird's-Foot Trefoil *Lotus corniculatus*	2	5
Black Medick *Medicago lupulina*	2	4
Meadow Vetchling *Lathyrus pratensis*	2	4
Sallow *Salix caprea/cinerea*	2	3
Common Vetch *Vicia sativa* agg.	2	2
unidentified garden flower (tall purple spikes)	2	2
Common Sowthistle *Sonchus oleraceus*	2	2
Teasel *Dipsacus fullonum*	2	2
Burdock *Arctium minus* agg.	2	2
Catmint *Nepeta cataria*	2	2
Hebe *Hebe* sp.	2	2
Marguerite	2	2

Petunia *Petunia* sp.	2	2
Hydrangea *Hydrangea* sp.	2	2
Welted Thistle *Carduus acanthoides*	2	2
Corn Chamomile *Anthemis arvensis*	1	4
Greater Stitchwort *Stellaria holostea*	1	3
Guelder Rose *Viburnum opulus*	1	2
Bush Vetch *Vicia sepium*	1	2
(unidentified white Crucifer)	1	2
Everlasting-flower *Helichrysum* sp.	1	2
Lesser Celandine *Ranunculus ficaria*	1	1
Green Alkanet *Pentaglottis sempervirens*	1	1
Broad-leaved Willowherb *Epilobium montanum*	1	1
Alsike Clover *Trifolium hybridum*	1	1
Clover (unspecified) *Trifolium* sp.	1	1
Herb Robert *Geranium robertianum*	1	1
Common Chickweed *Stellaria media*	1	1
Wild Radish *Raphanus raphanistrum*	1	1
Scentless Mayweed *Matricaria perforata*	1	1
Greater Knapweed *Centaurea scabiosa*	1	1
Hedge Mustard *Sisymbrium officinale*	1	1
Water Mint *Mentha aquatica*	1	1
Butterbur *Petasitis hybridus*	1	1
Common Spotted Orchid *Dactylorhiza fuchsii*	1	1
Common Hemp-Nettle *Galeopsis tetrahit* agg.	1	1
Ivy *Hedera helix*	1	1
Horse-radish *Armoracia rusticana*	1	1
Red Campion *Silene dioica*	1	1
Groundsel *Senecio vulgaris*	1	1
Forget-me-not *Myosotis* sp.	1	1
Shepherd's Purse *Capsella bursa-pastoris*	1	1
Fleabane *Pulicaria dysenterica*	1	1
Canadian Golden-Rod *Solidago canadensis*	1	1
Lilac *Syringa vulgaris*	1	1
Aubretia *Aubrieta columnae*	1	1
Wallflower *Cheiranthus cheiri*	1	1
Orange Ball Tree *Buddleja globosa*	1	1
Phlox *Phlox drummondii*	1	1
Pansy *Viola tricolor*	1	1
Marigold *Calendula officinalis*	1	1
Sweet Pea *Lathyrus odoratus*	1	1
garden Eupatorium *Eupatorium* sp.	1	1
(cultivated Crucifer)	1	1
Lobelia *Lobelia* sp. (?)	1	1
Geranium *Pelargonium* sp.	1	1
Rose *Rosa* sp.	1	1
Potato *Solanum tuberosum*	1	1
garden Cranesbill *Geranium* sp.	1	1
garden Antirrhinum *Antirrhinum majus*	1	1
garden Spiraea *Spiraea* sp.	1	1
African Marigold *Tagetes* sp.	1	1
Everlasting Pea *Lathyrus latifolius*	1	1
Dahlia *Dahlia* sp.	1	1
(blue Lantana-like flower)	1	1
(unidentified yellow garden Composite)	1	1
(unidentified white-flowered shrub)	1	1
(other unidentified garden flowers)	2	3

Table 3. USAGE OF NECTAR SOURCES BY EACH BUTTERFLY SPECIES

(numbers in brackets are the number of days on which the species was observed to use the source).

SMALL SKIPPER *Thymelicus sylvestris*.
Creeping Thistle (20), Red Clover (13), Tufted Vetch (10), Ragwort (8), Spear Thistle (7), Black Knapweed (6) White Clover (3), Bird's-Foot Trefoil (3), Rosebay Willowherb (2), Black Medick (1), Bramble (1), Meadow Vetchling (1), Greater Knapweed (1), Buddleia (1), Hawkweed etc. (1), Yarrow (1).
Total 16.

LARGE SKIPPER *Ochlodes venata*.
Creeping Thistle (13), Tufted Vetch (4), Red Clover (4), Black Knapweed (1).
Total 4.

BRIMSTONE *Gonepteryx rhamni*.
Buddleia (2), Spear Thistle (1), Creeping Thistle (1), Michaelmas Daisy (1), Herb Robert (1),
Total 5.

LARGE WHITE *Pieris brassicae*.
Buddleia (41), Dandelion (4), Creeping Thistle (4), Sweet Rocket (2), Red Clover (2), Rosebay Willowherb (2), wild Brassica (2), Wall Rocket? (1), Bramble (1), Spear Thistle (1), Black Knapweed (1), Michaelmas Daisy (1), Hogweed (1), Honesty (1), Bluebell (1), Fool's Parsley (1), Ragwort (1), Himalayan Balsam (1), Black Horehound? (1), Great Willowherb (1), Hedge Bindweed (1), Ivy (1), Lavender (1), Ice-plant (1), Geranium (1), Catmint (1), Garden Privet (1), Lilac (1), Petunia (1).
Total 29.

SMALL WHITE *Pieris rapae*.
Buddleia (26), Creeping Thistle (25), Bramble (19), Dandelion (17), Michaelmas Daisy (11), Lavender (9), Ragwort (8), wild Brassica (5), Rosebay Willowherb (4), Hawkweed etc.(4), Oxford Ragwort (3), Hedge Bindweed (3), Common Sowthistle (3), Himalayan Balsam (3), Green Alkanet (2), Spear Thistle (2), Black Knapweed (2), Hedge Mustard (2), Buttercup (2), Red Clover (2), Sweet Rocket (2), Corn Chamomile (2), Great Willowherb (2), Daisy (2), Hebe (2), unidentified yellow Crucifer (2), unidentified white Crucifer (2), unidentified garden flower (tall purple spikes) (2), cultivated Crucifer (1), Wild Radish (1), Cuckoo Flower (1), Garlic Mustard (1), Coltsfoot (1), Hawthorn (1), Common Chickweed (1), Horseradish (1), Red Campion (1), Groundsel (1), White Clover (1), Lucerne (1), Burdock (1), Black Horehound? (1), Comfrey (1), Red Valerian (1), Garden Privet (1), Rose (1), Potato (1), Pansy (1), Marigold (1), Sweet Pea (1), Marguerite (1), Lobelia? (1), French Marigold (1), garden Cranesbill (1), Ice-plant (1), Catmint (1), Antirrhinum (1), unidentified garden flowers (3), unidentified white-flowering shrub (1).
Total 59.

GREEN-VEINED WHITE *Pieris napi*.
Creeping Thistle (16), Dandelion (13), Bramble (11), Ragwort (8), Sweet Rocket (5), wild Brassica (5), Michaelmas Daisy (4), Buttercup (4), Cuckoo Flower (3), Greater Stitchwort (3), White Clover (3), Rosebay Willowherb (3), Fool's Parsley (2), Black Knapweed (2), Hedge Bindweed (2), Himalayan Balsam (2), Hawkweed etc. (2), Comfrey (1), Buddleia (1), Forget-me-not (1), Bluebell (1), Honesty (1), Hawthorn (1), Shepherd's Purse (1), Welted Thistle (1), Common Spotted Orchid (1), Lucerne (1), Black Horehound? (1), Common Hemp-Nettle (1), Garden Privet (1), Lavender (1), French Marigold (1), Hydrangea (1).
Total 33.

ORANGE-TIP *Anthocharis cardamines*.
Cuckoo Flower (5), Garlic Mustard (4), Sweet Rocket (4), Bush Vetch (2), Dandelion (2), Fool's Parsley (1), Common Vetch (1), Rosebay Willowherb (1), Bluebell (1), Honesty (1),

wild Brassica (1), unidentified yellow Crucifer (1), Aubretia (1).
Total 13.

SMALL COPPER *Lycaena phlaeas*.
Ragwort (16), Creeping Thistle (4), Hawkweed etc. (4), Dandelion (2), Michaelmas Daisy (2), Oxford Ragwort (1), Daisy (1), Ox-eye Daisy (1), Yarrow (1), Corn Chamomile (1), Scentless Mayweed (1), Water Mint (1), Red Clover (1), Buttercup (1).
Total 14.

COMMON BLUE *Polyommatus icarus*.
Ragwort (4), White Clover (4), Black Medick (3), Creeping Thistle (3), Bird's-Foot Trefoil (3), Tufted Vetch (2), Michaelmas Daisy (2), Red Clover (1), Alsike Clover (1), Buttercup (1), Yarrow (1), Bramble (1), Hawkweed etc.(1), Canadian Golden Rod (1), Rosebay Willowherb (1), Fleabane (1).
Total 16.

HOLLY BLUE *Celastrina argiolus*.
Buddleia (1), Michaelmas Daisy (1), Garden Privet (1), garden Eupatorium? (1).
Total 4.

RED ADMIRAL *Vanessa atalanta*.
Buddleia (51), Michaelmas Daisy (5), Creeping Thistle (3), Ragwort (3), Garden Privet (2), Bramble (1), Himalayan Balsam (1).
Total 7.

PAINTED LADY *Vanessa cardui*.
Buddleia (31), Creeping Thistle (13), Michaelmas Daisy (5), Black Knapweed (4), Ragwort (3), Oxford Ragwort (3), Red Clover (3), Bramble (3), Guelder Rose (2), Buttercup (2), White Clover (2), Hogweed (2), Garden Privet (2), Dandelion (1), Daisy (1), Sweet Rocket (1), Ox-eye Daisy (1), Comfrey (1), Common Vetch (1), Hawkweed etc.(1). Orange Ball Tree (1), Corn Chamomile (1), Heather (1), French Marigold (1), Ice-plant (1).
Total 25.

SMALL TORTOISESHELL *Aglais urticae*.
Buddleia (69), Creeping Thistle (33), Dandelion (30), Ragwort (23), Michaelmas Daisy (15), Ice-plant (11), Garden Privet (10), Hawkweed etc.(5), Black Knapweed (4), French Marigold (4), Coltsfoot (3), Lavender (3), Hogweed (2), Sallow (2), Bramble (2), Lucerne (2), Red Valerian (2), garden Everlasting Flower (2), Cuckoo Flower (1), Oxford Ragwort (1), Sweet Rocket (1), Red Clover (1), White Clover (1), Tufted Vetch (1), Butterbur (1), Heather (1), Phlox (1), wild Brassica (1), Fool's Parsley (1), Welted Thistle (1), Burdock (1), Common Sowthistle (1), garden Spiraea (1), Hydrangea (1), African Marigold (1), Everlasting Pea (1), Hebe (1), Dahlia (1), yellow garden Composite (1), unidentified garden flower (tall purple spikes) (1), blue Lantana-like flower (1), unidentified garden flower (1).
Total 42.

PEACOCK *Inachis io*.
Buddleia (30), Dandelion (15), Creeping Thistle (11), Ragwort (6), Michaelmas Daisy (3), Bramble (3), Red Clover (2), Lesser Celandine (1), Sweet Rocket (1), Sallow (1), Coltsfoot (1), Hawthorn (1), Buttercup (1), Fool's Parsley (1), Hawkweed etc.(1), Hogweed (1), Lucerne (1), Black Knapweed (1), Teasel (1), Red Valerian (1), Garden Privet (1), Petunia (1).
Total 22.

COMMA *Polygonia c-album*.
Buddleia (15), Michaelmas Daisy (8), Creeping Thistle (4), Ice-plant (3), Bramble (2), Bramble (fruit)(1), Ragwort (1), Cow Parsley (1), Coltsfoot (1).
Total 8.

SPECKLED WOOD *Pararge aegeria*.
Bramble (2), Ragwort (2), Buddleia (1), Creeping Thistle (1), Michaelmas Daisy (1), Himalayan Balsam (1).
Total 6.

WALL *Lasiommata megera*.
Oxford Ragwort (2), Buttercup (1), Hawkweed etc.(1), Ragwort (1), Heather (1), Buddleia (1), Creeping Thistle (1), Wallflower (1).
Total 8.

GATEKEEPER *Pyronia tithonus*.
Creeping Thistle (2), Ox-eye Daisy (1), Ragwort (1), Buddleia (1), Bramble (1).
Total 5.

MEADOW BROWN *Maniola jurtina*.
Creeping Thistle (20), Ragwort (14), Tufted Vetch (5), Black Knapweed (5), White Clover (4), Bramble (3), Red Clover (2), Clover (unspecified) (1), Hogweed (1), Hawkweed etc.(1), Spear Thistle (1), Meadow Vetchling (1), Michaelmas Daisy (1), Marguerite (1).
Total 14.

REFERENCES.

Baker, R.R. (1969). The evolution of the migratory habitat in butterflies. *Journal of Animal Ecology*, **38**: 703-746.

Blackie, J.E.H. (1946). Butterflies of a Manchester Housing Estate. North Western Naturalist, **XXI**: 185-187.

Courtney, S.P. (1980). Studies on the Biology of Butterflies *Anthocharis cardamines* (L.) and *Pieris napi* (L.) in relation to speciation in the Pierinae. Ph.D. thesis, University of Durham.

Courtney, S.P. (1981). Coevolution of pierid butterflies and their cruciferous hostplants. III. *Anthocharis cardamines* (L.) survival, development and oviposition on different hostplants. *Oecologia*, 51: 91-96.

Courtney, S.P. (1982A). Coevolution of pierid butterflies and their cruciferous hostplants. IV. Crucifer apparency and *Anthocharis cardamines* (L.) oviposition. *Oecologia*, 52: 258-265.

Courtney, S.P. (1982b). Coevolution of pierid butterflies and their cruciferous hostplants. V. Habitat selection, structure and speciation. *Oecologia*, 54: 101-107.

Dennis, R.L.H. (1982). Observation on habitats and dispersion made from oviposition markers in north Cheshire *Anthocharis cardamines* (L.) (Lepidoptera: Pieridae). *Entomologist's Gazette*, **33**: 151-159.

Dennis, R.L.H. (1985a). Small plants attract attention! Choice of egg-laying sites in the green-veined white butterfly *Artogeia napi* (L) (Lep: Pieridae). *The Bulletin of the amateur Entomologists' Society*, **44**: 77-82.

Dennis, R.L.H. (1985b). Voltinism in British *Aglais urticae* (L) (Lep: Nymphalidae): variation in space and time. *Proceedings and Transactions of the British Entomological and Natural History Society*, **18**: 51-61.

Dennis, R.L.H. (1985c). *Polyommatus icarus* (Rottemburg) (Lepidoptera: Lycaenidae) on Brereton Heath in Cheshire: voltinism and switches in resource exploitation. *Entomologist's Gaz.* **36**: 175-179.

Dennis, R.L.H. (Ed.) (1992A). *The Ecology of Butterflies in Britain.* Oxford University Press.

Dennis, R.L.H. (1992B). The spread of the small skipper butterfly in North Cheshire apparently by '*Ho[l]cus* pocus'. *Butterfly Conservation News, East Cheshire & Peak District Branch*, **20**:5.

Dennis, R.L.H. (1993). *Butterflies and climate change.* Manchester University Press.

Dennis, R.L.H. & Shreeve, T.G. (1996). Butterflies on British and Irish offshore islands. Gem Publishing Co.

Dennis, R.L.H. & Williams, W.R. (1987). Mate-location behaviour of the large skipper butterfly *Ochlodes venata*. Flexible strategies and spatial components. *Journal of the Lepidopterists' Society*, **41**: 45-64.

Emmet, A.M. & Heath, J. (Eds) (1989). *The Moths and Butterflies of Great Britain and Ireland* **7**(1). Colchester.

Fiedler, K. (1990). Effects of larval diet on myrmecophilous qualities of *Polyommatus icarus* caterpillars (Lepidoptera: Lycaenidae). *Oecologia*, **83**: 284-287.

Fiedler, K. & Holldobler, B. (1992). Ants and *Polyommatus icarus* immatures (Lycaenidae) - sex-related development benefits and costs of ant attendance. *Oecologia*, **91**: 468-473.

Findlay, J.D, & Tilley, R.J.T. (1995). Overwintering of larvae of the Speckled Wood butterfly *Pararge aegeria* (L.)(Lepidoptera: Satyridae), in southern Britain. *Entomologist's Gaz.* **46**: 173-176.

Fitter, R., Fitter, A. & Blamey, M. (1974). *The wild flowers of Britain and Northern Europe*. Collins, London.

Garland, S.P. (1981). Butterflies of the Sheffield area. Sorby Natural History Society & Sheffield City Museums.

Geographers' A-Z Map Company Ltd. (1993). *New Greater Manchester A-Z Street Atlas.* London.

Hardy, P.B. (1994). Monitoring of selected butterfly species and their hostplant-habitats by 100 m^2 units. *Entomologist's Gaz.* **45**: 159-166.

Hardy, P.B. (1995). Scarcity of Vanessid butterflies. *Ent. Rec. J. Var.* **107**: 257.

Hardy, P.B. & Dennis, R.L.H. (1997). Butterfly range-extension into Greater Manchester: the role of climate change and habitat patches. *Urban Nature* **3(1)**: 6-8.

Hardy, P.B, Hind, S.H. & Dennis, R.L.H. (1993). Range extension and distribution-infilling among selected butterfly species in north-west England: evidence for inter-habitat movements. *Entomologist's Gaz.* **44**: 247-255.

Highways Agency (1995). M63 junctions 6 to 9 widening, pre-public inquiry statement.

Huxley, A. (1967). *Mountain flowers of Europe*. Blandford, Dorset.

Newman, L.H. (1967). *Create a Butterfly Garden*. J. Baker, London.

Owen, D.F. (1949). The Macrolepidoptera of the Moorgate, London, bombed sites. *Entomologist* **82**: 59-62.

Pollard, E. & Greatorex-Davies, J.N. (1997). Flight-periods of the Small Heath butterfly, *Coenonympha pamphilus* (Linnaeus) (Lepidoptera: Nymphalidae, Satyrinae) on chalk downs and in woodland in southern England. *Entomologist's Gazette*, **48**: 3-7.

Pollard, E., Hall, M.L. & Bibby, T.J. (1986). *Monitoring the abundance of butterflies 1976 - 1985.* Nature Conservancy Council.

Rothschild, M. & Farrell, C. (1983). *The Butterfly Gardener.* Rainbird, London.

Rutherford, C.I. (1983). Butterflies in Cheshire 1961 to 1982. Lancashire & Cheshire Entomological Society.

Sanders, E. (1939). *A butterfly book for the pocket.* Oxford University Press.

Shreeve, T.G. (1984). Habitat selection, mate-location, and microclimatic restraints on the activity of the speckled wood butterfly *Pararge aegeria*. *Oikos*, **42**:371-377.

Shreeve, T.G.(1986). The effect of weather on the life cycle of the speckled wood butterfly, *Pararge aegeria. Ecological Entomology*, 11: 325-352

Shreeve, T.G., Dennis, R.L.H. & Williams, W.R. (1996). Uniformity of wing spotting of *Maniola jurtina* (L.) in relation to environmental heterogeneity (Lepidoptera: Satyrinae). *Nota lepidopterologica*, **18**: 77-92.

Smith, C.J. (1995). The scarcity of Vanessid butterflies. *Ent. Rec. J. Var.* **107**: 146.

Thomas, C.D. (1984). Oviposition and egg load assessment by *Anthocharis cardamines* (L.) (Lep: Pieridae). *Entomologist's Gazette,* 35: 145-148.

Vickery, M. (1990). National garden butterfly survey - Preliminary report. British Butterfly Conservation Society News **44**: 37-39.

Vickery, M. (1991a). National garden butterfly survey 1990 report part 1. Butterfly Conservation News **47**: 32-38.

Vickery, M. (1991b). National garden butterfly survey 1990 report - part 2. Butterfly Conservation News **48**: 26-30.

Vickery, M. (1992). Garden butterfly survey 1991 report. Butterfly Conservation News **50**: 42-49.

Vickery, M. (1993). Garden butterfly survey 1992. Butterfly Conservation News **53**: 50-59.

Vickery, M. (1994). Butterfly Conservation national garden butterfly survey 1993. Butterfly Conservation News **56**: 45-50.

Vickery, M. (1995). Butterfly Conservation national garden butterfly survey 1994. Butterfly Conservation News **59**: 26-28.

Vickery, M. (1996). The garden butterfly survey 1995. Butterfly Conservation News **62**: 25-27.

Whitehead, R.W. (1986). Butterflies of Merseyside. British Butterfly Conservation Society, Merseyside Branch.

Wiltshire, E.P. (1997). *Entomologist*, **116**: 58-65.

ENVIRONMENTS - WHOLE AREA

Fig.1. GREATER MANCHESTER - ADMINISTRATIVE DISTRICTS

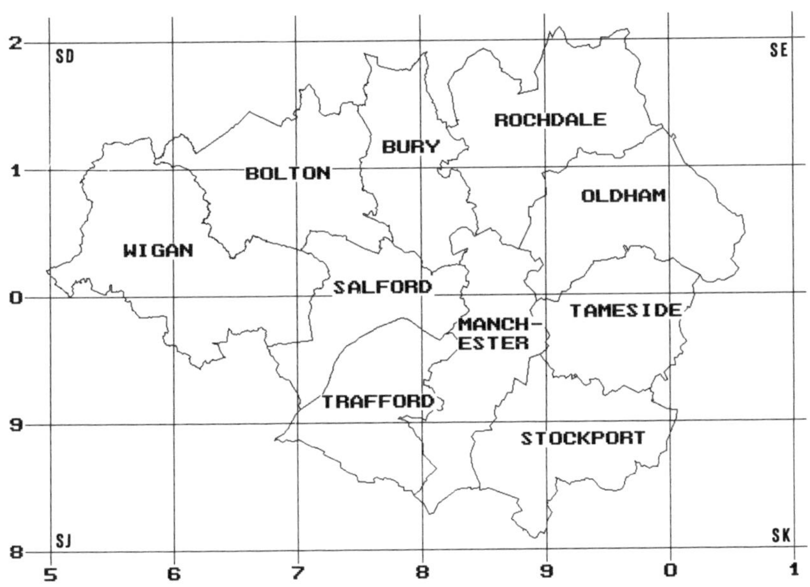

Fig.2. CANALS, RIVERS AND STANDING WATER

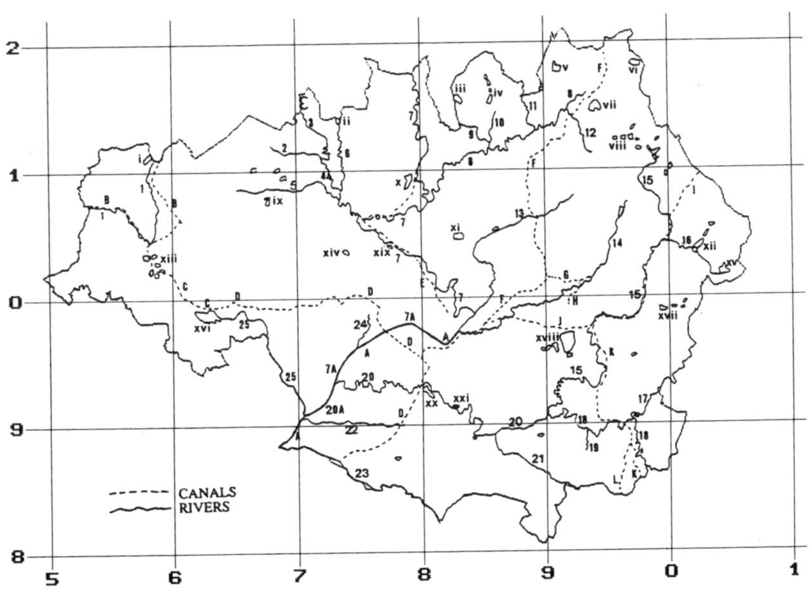

ENVIRONMENTS - WHOLE AREA

Fig.3. RAILWAYS AND MAIN ROADS

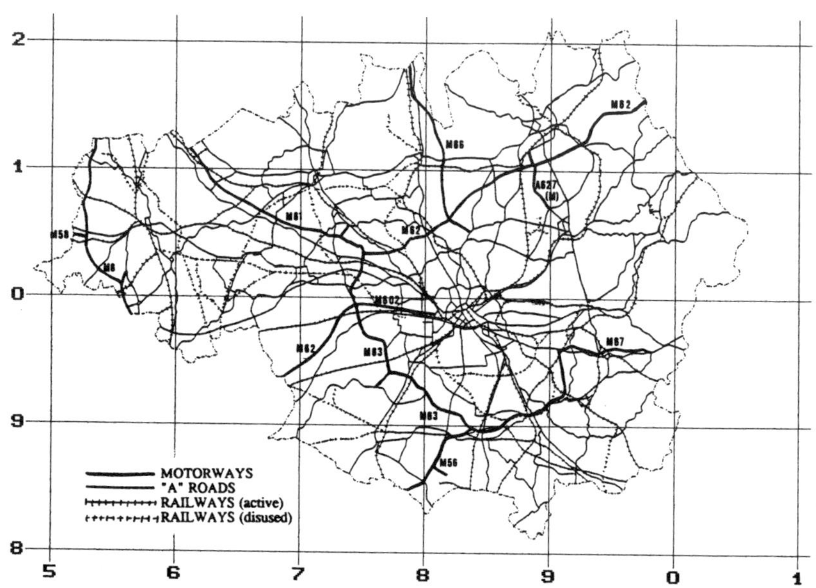

Fig.4. NATURE RESERVES, PARKS, OTHER NOTABLE WILDLIFE SITES

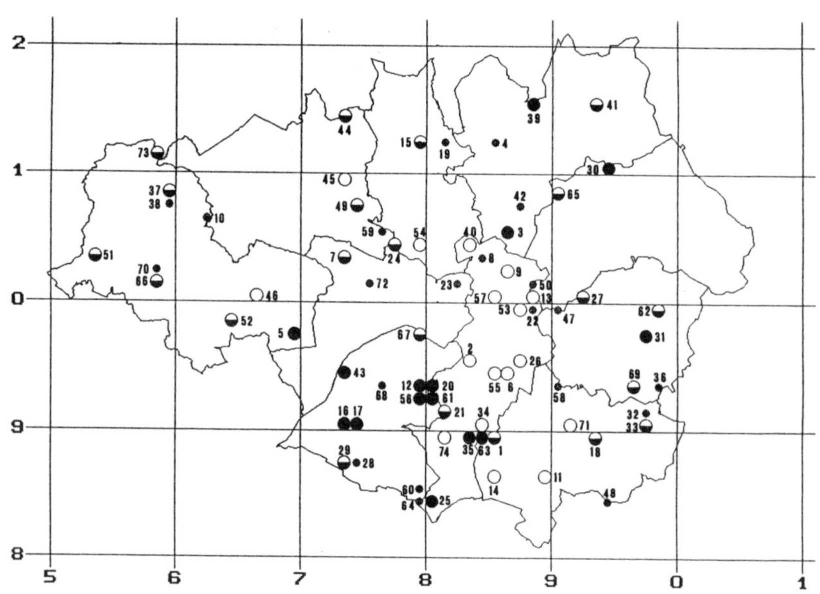

RIVERS

1. Douglas
2. Dean Brook/Astley Brook
3. Eagley Brook
4. Tonge
5. Croal
6. Bradshaw Brook
7. Irwell

7A. Manchester Ship Canal (canalisation of Irwell)

8. Roch
9. Cheesden Brook
10. Ashworth
11. Spodden
12. Beal
13. Irk
14. Medlock
15. Tame
16. Chew Brook
17. Etherow
18. Goyt
19. Poise Brook
20. Mersey

20A. Manchester Ship Canal (canalisation of Mersey + Irwell)

21. Lady Brook/Micker Brook
22. Timperley Brook/Sinderland Brook/Red Brook
23. Bollin
24. Salteye Brook
25. Pennington Brook/Glaze Brook

CANALS

A. Manchester Ship Canal (canalisation of rivers Irwell + Mersey)
B. Leeds & Liverpool Canal
C. Leeds & Liverpool Canal (Leigh Branch)
D. Bridgewater Canal
E. Manchester, Bolton & Bury Canal
F. Rochdale Canal
G. Fairbottom Branch Canal
H. Hollinwood Branch Canal
I. Huddersfield Narrow Canal
J. Manchester & Ashton-under-Lyne Canal
K. Peak Forest Canal
L. Macclesfield Canal

LAKES AND RESERVOIRS

i. Worthington Lakes
ii. Jumbles Reservoir
iii. Ashworth Moor Reservoir
iv. Naden & Greenbooth Reservoirs
v. Watergrove Reservoir
vi. Blackstone Edge Rerservoir
vii. Hollingworth Lake
viii. complex including Ogden, Piethorn, Rooden, Dowry and other Reservoirs
ix. Rumworth Lodge Reservoir
x. Manchester, Bolton & Bury reservoir
xi. Heaton Park Reservoir
xii. Dovestone Reservoir
xiii. Wigan Flashes including Scotsman's Flash, Pearson's Flash
xiv. Blackleach Reservoir
xv. Chew Reservoir
xvi. Pennington Flash
xvii. Gorton & Audenshaw Reservoirs
xix. Clifton Water Park
xx. Sale Water Park
xxi. Chorlton Water Park

DESIGNATED NATURE RESERVES (BLACK SYMBOLS), COUNTRY PARKS (BLACK/WHITE SYMBOLS), URBAN PARKS WITH WILDLIFE INTEREST (WHITE SYMBOLS) AND OTHER NOTABLE WILDLIFE SITES WITH PUBLIC ACCESS (WOODLANDS, WETLANDS) (SMALL GREY SYMBOLS) (with, where appropriate, species account, or introductory section, under which they are mentioned)

1. Abney Hall Park SJ8589
2. Alexandra Park SJ8395
3. Alkrington Woods SD8604/8605 (*A.aglaja*)
4. Ashworth Valley SD8512
5. Astley Moss SJ6997
6. Birchfields Park SJ8694 (species richness & distribution)
7. Blackleach Country Park SD7303
8. Blackley Forest SD8403
9. Boggart Hole Clough SD8602
10. Borsdane Wood SD6206
11. Bramhall Park SJ8986
12. Broad Ees Dole (Sale Water Park) SJ7993/8093 (distribution changes)
13. Brookdale Park SD8800/8900
14. Bruntwood Park SJ8586/8587
15. Burrs Country Park SD7912
16. Carrington Moss (Birchmoss Covert, Brookheys Covert, Hogswood Covert) SJ7490
17. Carrington Moss (Dark Lane) SJ7390 (*C.rubi*)
18. Chadkirk Country Park SJ9389/9489 (*S.w-album, V.cardui, M.jurtina*)
19. Chesham Woods, Bury SD8112
20. Chorlton Ees SJ8093 (distribution changes, *P.aegeria*)
21. Chorlton Water Park SJ8191/8291
22. Clayton Vale SJ8899 (*P.icarus*)
23. The Cliff/Kersal Vale SD8201
24. Clifton Country Park SD7704
25. Cotterill Clough SJ8083
26. Crowcroft Park SJ8795 (*P.c-album*)
27. Daisy Nook SD9200
28. Dunham New Park SJ7487/7587
29. Dunham Park SJ7387
30. Dunwood Park SD9410
31. East Wood SJ9797
32. Ernocroft Wood SJ9791
33. Etherow Country Park, Compstall SJ9690/9790 (*S.w-album*)
34. Fletcher Moss Park SJ8490
35. Gatley Carrs SJ8389
36. Great Wood/Hurst Clough SJ9893
37. Haigh Hall & Country Park SD5908 (*I.io, P.tithonus*)
38. Haigh Plantations SD5907
39. Healey Dell SD8815
40. Heaton Park SD8304
41. Hollingworth Lake Country Park SD9315
42. Hopwood Clough SD8707
43. Jack Lane, Flixton SJ7393
44. Jumbles Country Park SD7314
45. Leverhulme Park SD7309
46. Lilford Park SD6600 (*V.atalanta*)
47. Medlock Vale SJ9099
48. Middle Wood SJ9484
49. Moses Gate Country Park SD7406/7407 (*V.atalanta, A.urticae*)
50. Moston Fairway SD8801
51. Orrell Water Park SD5303
52. Pennington Flash Country Park SJ6398/6399/6498/6499 (*P.icarus*)
53. Philips Park, Bradford SJ8799
54. Philips Park, Prestwich SD7904
55. Platt Fields Park SJ8594
56. Priory, Sale SJ7992/8092 (distribution changes, *P.brassicae, A.cardamines, A.urticae, P.aegeria*)
57. Queen's Park, Harpurhey SD8500
58. Reddish Vale (Tame Valley) SJ9093 (*P.napi*)
59. Ringley Wood SD7605 (*V.atalanta, P.c-album*)
60. Rossmill (Bollin Valley) SJ7884 (*A.cardamines*)
61. Sale Water Park SJ8092 (distribution changes, *G.rhamni, P.aegeria*)
62. Stalybridge Country Park (Walkerwood, Brushes etc.) SD9799/9899
63. Stenner Woods SJ8490
64. Sunbank Wood SJ7984
65. Tandle Hill Country Park SD9008
66. Three Sisters Country Park SD5801
67. Trafford Ecology Park SJ7997
68. Urmston Meadows SJ7693/7694
69. Werneth Low Country Park SJ9693
70. Wigan Flashes SD5802/5803
71. Woodbank Park SJ9190
72. Worsley Woods SD7501 (*P.aegeria*)
73. Worthington Lakes SD5810/5811
74. Wythenshawe Park SJ8189

WHOLE AREA

Fig.5. OTHER SITES MENTIONED IN THE TEXT

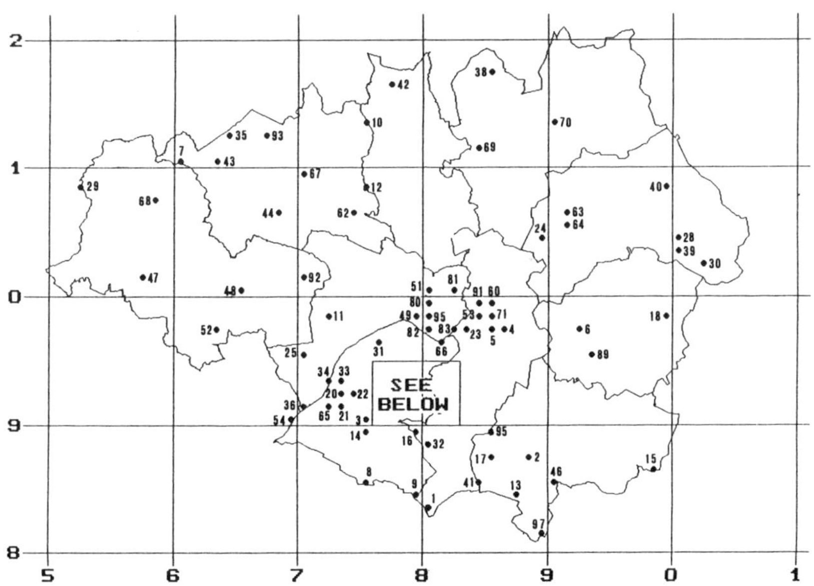

7 km x 5 km ZONE

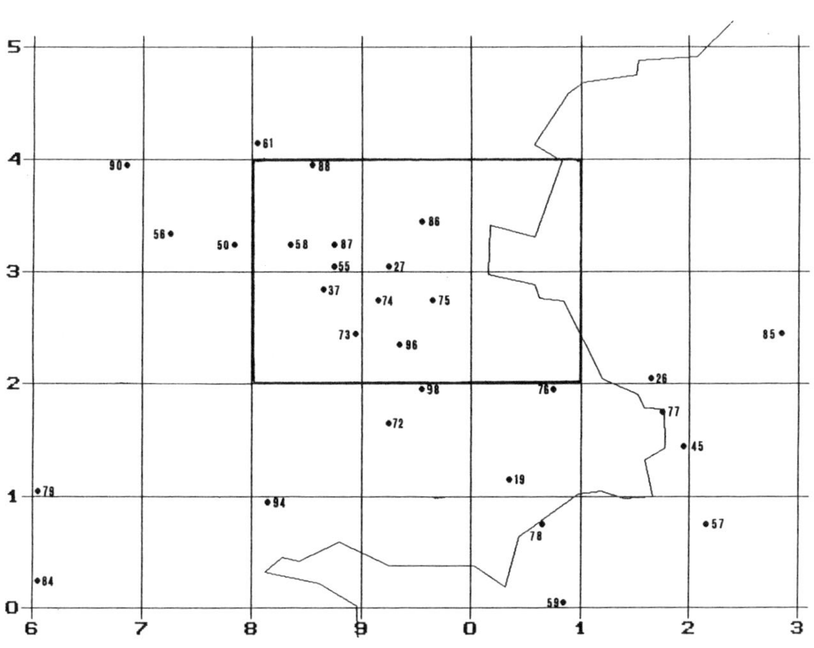

OTHER SITES MENTIONED IN TEXT (with species account, or introductory section, under which they are mentioned). Private gardens are not included.

1. A.538 road cutting, near airport SJ8083 (*T.sylvestris, O.venata, P.tithonus, M.jurtina, C. pamphilus*)
2. Adswood tip SJ8887 (*C.croceus, C.pamphilus)*
3. Altrincham sewage farm (railway near) SJ7590 (*Q.quercus*)
4. Ardwick, disused railway yard SJ8697 (*T.sylvestris, M.jurtina*)
5. Ardwick, Temperance Street SJ8597 (*Z.filipendulae*)
6. Ashton-under-Lyne sewage farm SJ9297 (*P.tithonus*)
7. Blackrod SD6010+ (*I.io*)
8. Bollin Valley, Bowdon SJ7585 (*A.cardamines*)
9. Bollin Valley, woods near Rossmill/Pigley Stairs) SJ7884/7983/7984 (*A.cardamines, P.aegeria*)
10. Bolton, hills east of SJ7513 (*A.urticae*)
11. Botany Bay Wood SJ7298/7398/7399 (distribution changes, *C. rubi, I.io*)
12. Bradley Fold SD7508 (*A.urticae, I.io*)
13. Bramhall, former military site SJ8784 (*I.io*)
14. Broadheath (Sinderland Road & Dairyhouse Farm) SJ7589/7590 (*P.rapae, V.atalanta, V.cardui*)
15. Brookbottom SJ9886 (*C.rubi*)
16. Brooklands estate, wood SJ7989 (distribution changes; *P.aegeria*)
17. Bruntwood hay meadow (destroyed) SJ8587 (conservation)
18. Brushes, Cock Wood SJ9998 (*C.rubi*)
19. Budworth Road sports field, Sale SJ8091 (*A.urticae*)
20. Carrington, horse field SJ7392 (*P.tithonus*)
21. Carrington Moss, disused railway & nearby SJ7391 (*L.megera, P.tithonus*)
22. Carrington, Shell pond SJ7492 (*Z.filipendulae*)
23. Castlefield SJ8397 (*P.icarus*)
24. Chadderton sewage farm SD8904 (*P.napi*)
25. Chat Moss SJ7095 (distribution changes)
26. Chorlton tip SJ8192 (*A.aglaja*)
27. Crossford sports field, Sale SJ7993 (conservation; *A.cardamines, A.urticae, I.io, Z.lonicerae*)
28. Dick Clough, Tunstead SE0004 (*C.rubi*)
29. Douglas Valley, Gathurst SD5208 (*A.cardamines*)
30. Dovestone Reservoir, field east of SE0202 (*I.io*)
31. Dumplington (largely destroyed) SJ7696/7796 (conservation)
32. Fairywell Brook SJ8088 (*P.aegeria*)
33. Flixton fly-ash tip SJ7393 (*P.icarus, A.aglaja, P.tithonus, Z.filipendulae*)
34. Flixton, wood nr. Manchester ship canal SJ7293 (*P.tithonus*)
35. Foxholes SD6412 (*C.rubi*)
36. Glaze/Manchester ship canal confluence, wood near SJ7091 (*A.cardamines*)
37. Glebelands Road sports field, Sale (destroyed) SJ7892 (conservation)
38. Greenbooth reservoir SD8517 (*C.pamphilus*)
39. Greenfield, 1937 social club SE0003 (*M.jurtina*)
40. Harrop Edge SD9908 (*S.w-album*)
41. Heald Green, railway cutting SJ8485 (*T.sylvestris, C.croceus, P.tithonus, C.pamphilus*)
42. Holcombe Moor SD7716 (*O.venata, C.rubi, M.jurtina*)
43. Horwich, former railway sidings nr. Red Moss SD6310 (*P.icarus, Z.filipendulae*)
44. Hunger Hill SD6806 (*T.sylvestris*)
45. Kenworthy SJ8191/8291 (*V.atalanta*)
46. Ladybrook Valley SJ9085 (*O.venata*)
47. Land Gate, former tip SD5701 (*P.icarus*)
48. Leigh (waste ground) SD6500 (*V.atalanta*)
49. Little Bolton, mineral railway SJ7998 (*V.atalanta*)
50. Little Ees Lane, Ashton-on-Mersey SJ7793 (*A.urticae*)
51. Littleton Road playing fields, Salford SD8000/8001/8100 (*A.urticae*)

52. Lowton Common SJ6397 (*A.urticae*)
53. Manchester city centre SJ8498 (species richness & distribution, *V.atalanta*)
54. Manchester Ship Canal, Warburton SJ6990 (*P.napi*)
55. Mersey bank, Stretford SJ7892/7893 (conservation; *O.venata, C.croceus, A.cardamines, L.phlaeas, P.icarus, V.atalanta, L.megera, P.tithonus, M.jurtina*)
56. Mersey valley, Ashton-on-Mersey golf course SJ7793 (*A.cardamines*)
57. Mersey Valley, Northenden SJ8290 (*A.cardamines*)
58. Mersey Valley, site near Stretford sewage farm SJ7893 (*T.sylvestris, P.c-album, P.aegeria, P.tithonus*)
59. Nan Nook Wood SJ8090 (*P.aegeria*)
60. New Allen Street SJ8599 (*M.jurtina*)
61. Newcroft tree nursery SJ7894 (*A.cardamines*)
62. Nob End SD7406 (*P.icarus*)
63. Oldham (open site) SD9106 (*V.cardui*)
64. Oldham (waste) SD9105 (*V.cardui*)
65. Partington, site near B.P.works SJ7291 (distribution changes; *O.venata, P.icarus, P.c-album, P.tithonus*)
66. Pomona docks SJ8196 (*P.icarus*)
67. Queen's Park, Bolton SD7009 (*L.boeticus*)
68. Red Rocks disused railway, north of Wigan SD5807/8 (*P.tithonus*)
69. Roch Valley, Hooley Bridge SD8411 (*A.cardamines*)
70. Roch Valley, Rochdale SD9013 (*C.pamphilus*)
71. Rochdale Canal, Ancoats SJ8598 (*P.rapae*)
72. Sale, abandoned tennis court SJ7991 (*A.urticae*)
73. Sale, alley behind Harley Road SJ7892 (*P.rapae*)
74. Sale, alley behind Meadows Road SJ7992 (*P.brassicae*)
75. Sale, former Bethell's tip SJ7992 (distribution changes; *P.icarus*)
76. Sale golf club SJ8091 (*C.argiolus*)
77. Sale, limestone "raft" by electricity station SJ8191 (*P.icarus*)
78. Sale/Northern Moor, old nurseries (destroyed) SJ8090 (conservation)
79. Sale, "Racecourse" estate SJ7691 (*G.rhamni*)
80. Salford (Fitzwarren Street) SJ8098 (*V.atalanta*)
81. Salford (Great Clowes Street) SD8200 (*L.phlaeas*)
82. Salford Quays SJ8097 (Manchester's environments; *V.atalanta*)
83. Salford (waste land) SJ8297 (*L.phlaeas*)
84. Sinderland Brook SJ7590/7690 (*P.icarus, P.tithonus*)
85. Southern Cemetery SJ8292 (*L.phlaeas*)
86. Stretford cabbage fields, Mosley Acre Farm SJ7993 (*P.brassicae, P.rapae, A.urticae*)
87. Stretford, New Manor Farm SJ7893 (*P.brassicae, V.cardui*)
88. Stretford tip SJ7893/4 (distribution changes; *T.sylvestris, P.icarus, V.atalanta, M.jurtina*)
89. Tame Valley, Denton SJ9395 (*M.jurtina*)
90. Urmston oxbow SJ7693 (*I.io*)
91. Victoria station, waste ground near (Scotland & Angel Meadow) SJ8499 (*M.jurtina*)
92. Walkden - Leigh disused railway SD6901-7201 (*P.aegeria*)
93. Walker Fold SD6712 (*P.napi*)
94. Walton Road Park, Sale SJ7890 (*P.aegeria*)
95. Waterside Hotel, Didsbury SJ8589 (*N.antiopa*)
95. Weaste cemetery SJ8097/8 (*L.phlaeas*)
96. Winstanley Road allotments, Sale SJ7992 (*P.rapae, P.napi*)
97. Woodford SJ8981 (*A.cardamines*)
98. Worthington Park SJ7991/2 (*P.aegeria*)

URBAN COVER - WHOLE AREA

Fig.6. URBAN COVER

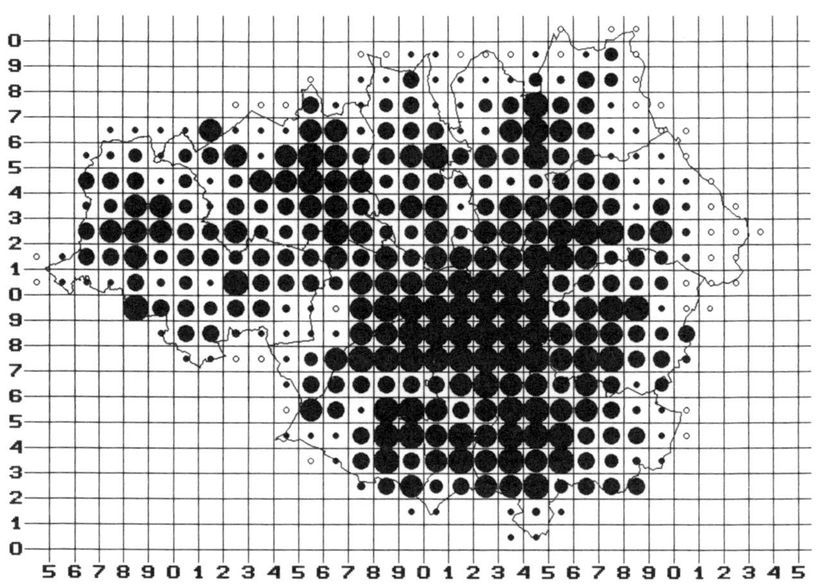

- Nil cover
- 1-5% cover
- 6-10% cover
- 11-20% cover
- 21-50% cover
- 51-75% cover
- 76-100% cover

RECORDING COVERAGE - WHOLE AREA

Symbols graded from 1 visit (smallest size) to 100+ visits. Crosses, no visits.

Fig.7. RECORDING COVERAGE BY TETRADS

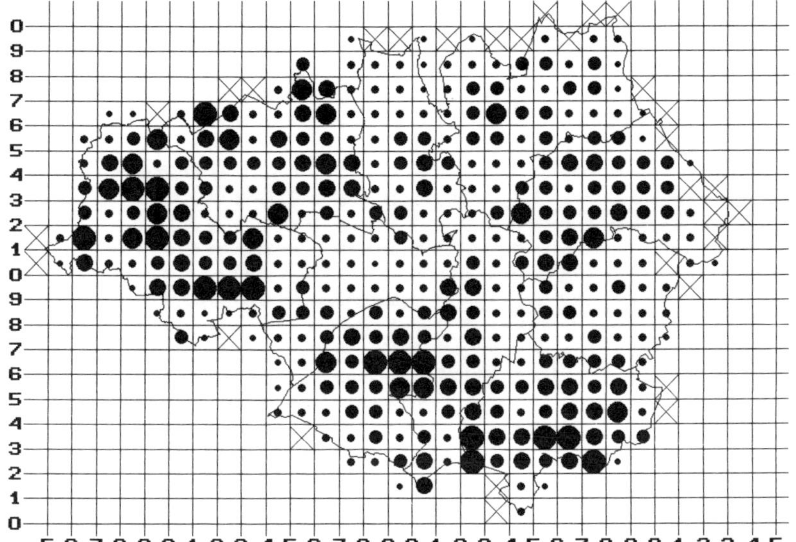

Fig.8 RECORDING COVERAGE BY 1KM SQUARES

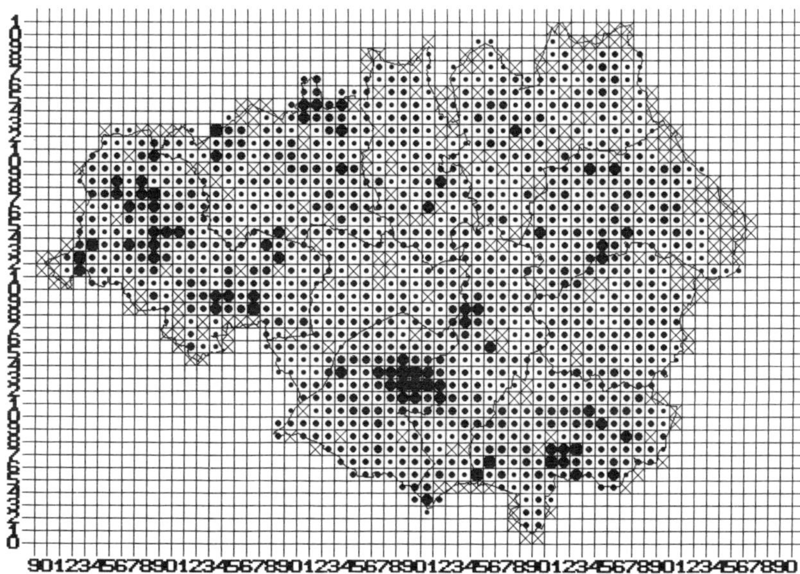

BUTTERFLY DISTRIBUTIONS, 1980-1997 - WHOLE AREA

Fig.9. RECORDS FOR ALL BUTTERFLY SPECIES PER TETRAD

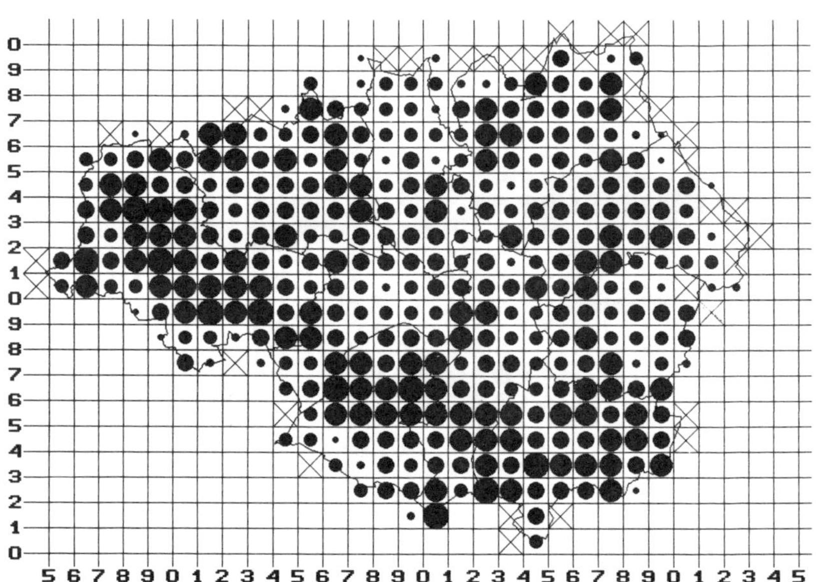

NUMBERS OF SPECIES RECORDED IN EACH TETRAD

																					0		0	0					
													1	0	0	2	0	0	0	0	10	0	3	6					
											7		4	7	7	7	2	2	5	15	14	7	15	0					
								0	0	2	17	14	9	6	8	2	11	17	13	14	10	12	15	0	0				
			0	1	0	3	15	14	7	6	14	16	10	5	8	7	8	15	17	13	12	10	8	3	1	0			
		5	6	10	17	11	15	14	10	19	8	16	6	3	13	2	8	15	14	8	14	8	15	11	1	0			
		8	16	17	12	12	9	4	13	14	14	16	15	9	10	15	11	7	4	9	9	12	12	11	10	11	1		
		14	18	20	21	17	14	7	10	14	13	14	17	13	9	15	4	10	8	13	12	12	12	14	12	8	0	0	
		13	6	15	18	17	12	7	11	17	6	9	11	14	12	10	5	7	16	14	11	11	15	11	15	14	2	0	0
0	11	21	12	17	21	17	10	17	14	9	12	16	12	13	12	11	9	11	4	8	11	15	15	8	6	9	7	0	
0	6	18	9	7	16	17	17	17	15	13	9	13	8	2	8	10	14	12	11	15	13	18	6	8	4	0	2	2	
				1	12	16	21	19	20	11	16	10	7	6	6	6	15	15	7	5	13	12	7	13	12	12	0		
					1	8	8	4	10	16	16	10	9	10	12	11	15	13	6	9	14	17	8	9	8	11			
						12	4	0	4	7	8	16	16	13	15	16	8	14	10	8	6	11	15	4	8	4			
										8	10	21	15	18	20	19	14	13	5	8	11	16	15	13	18				
										0	9	17	17	15	17	18	16	17	17	10	17	18	12	16	10	0			
										9	6	1	12	12	11	14	17	18	15	13	13	14	18	19	10	0			
											0	9	4	11	6	14	11	19	13	20	19	16	17	14	18				
													6	11	13	16	7	21	19	13	11	14	16	1					
															4	21			0	10	0								
																				0	6								

BUTTERFLY DISTRIBUTIONS, 1980-1997 - WHOLE AREA

Fig.10. Thymelicus sylvestris

Fig.11. Ochlodes venata

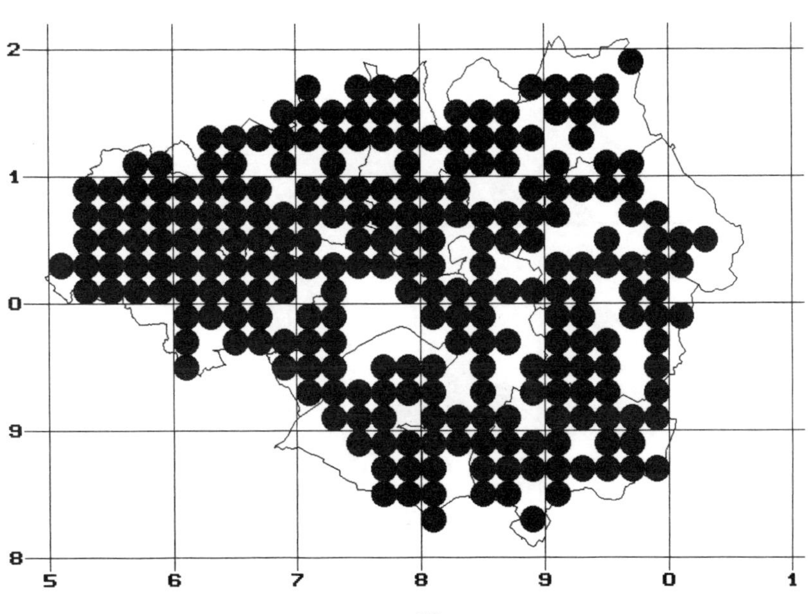

BUTTERFLY DISTRIBUTIONS, 1980-1997 - WHOLE AREA

Fig.12. Colias croceus

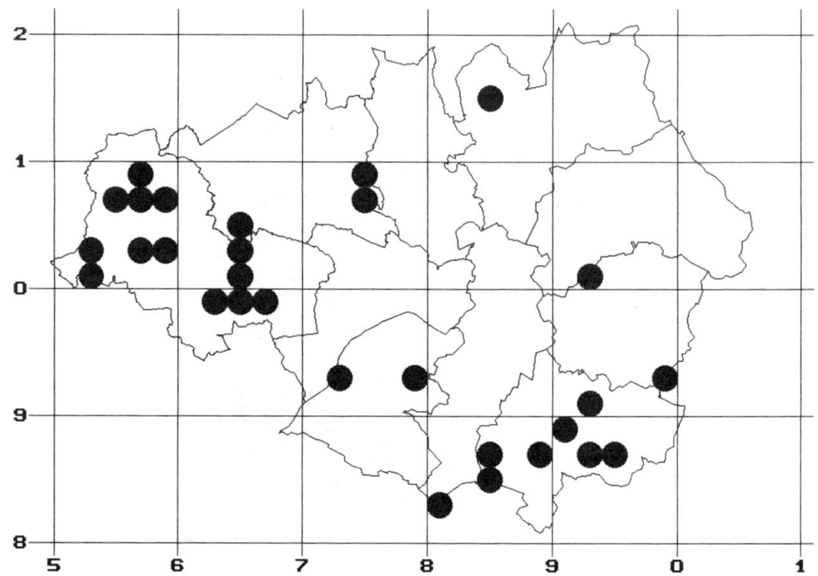

Fig.13. Gonepteryx rhamni

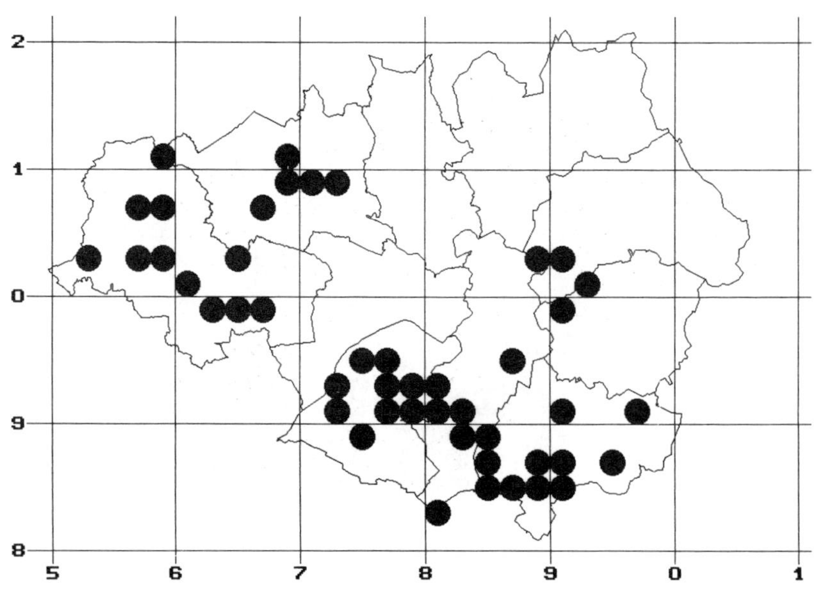

Fig.14. Pieris brassicae

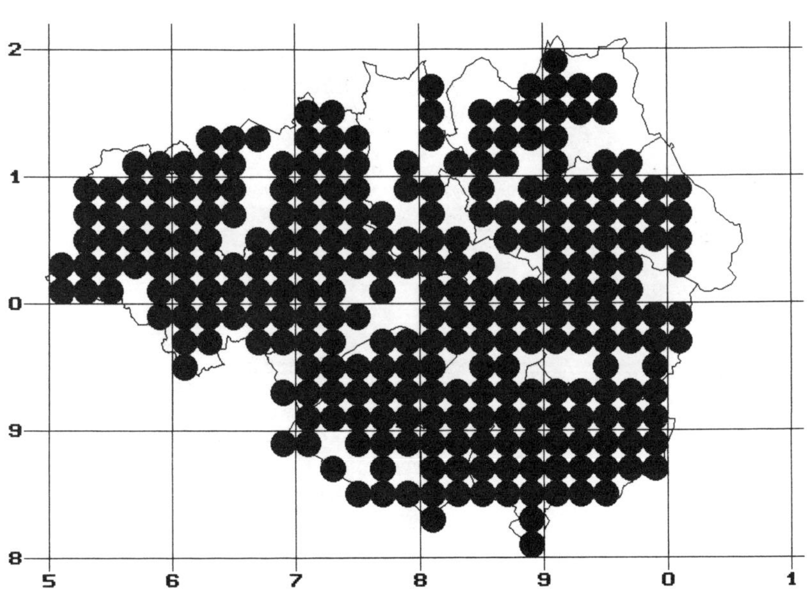

Fig.15. Pieris rapae

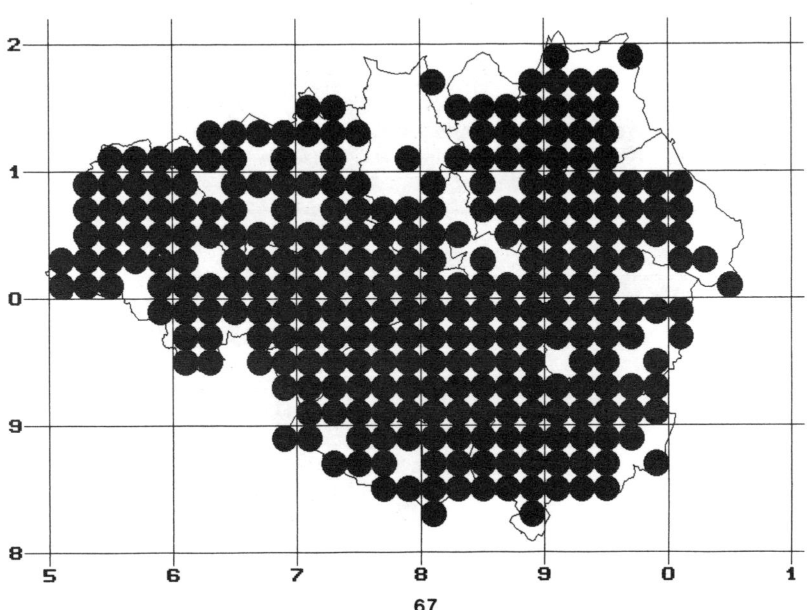

BUTTERFLY DISTRIBUTIONS, 1980-1997 - WHOLE AREA

Fig.16. Pieris napi

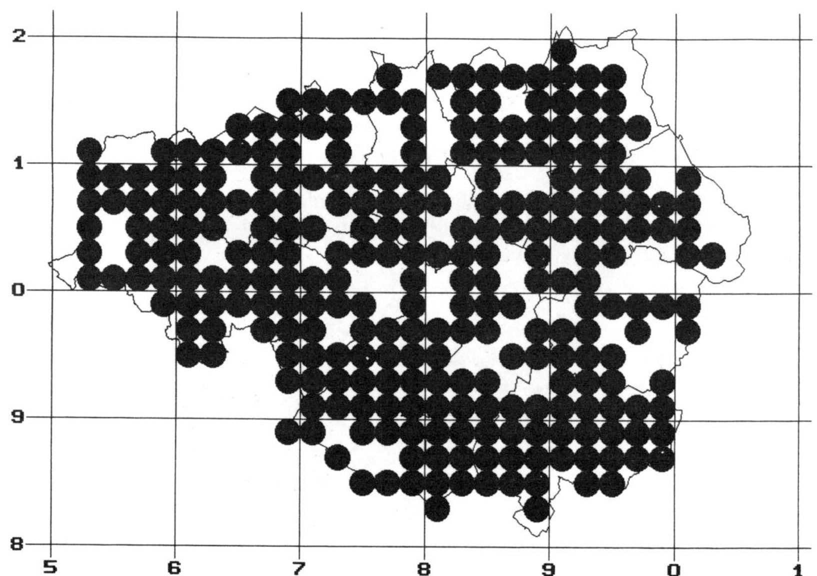

Fig.17. Anthocharis cardamines

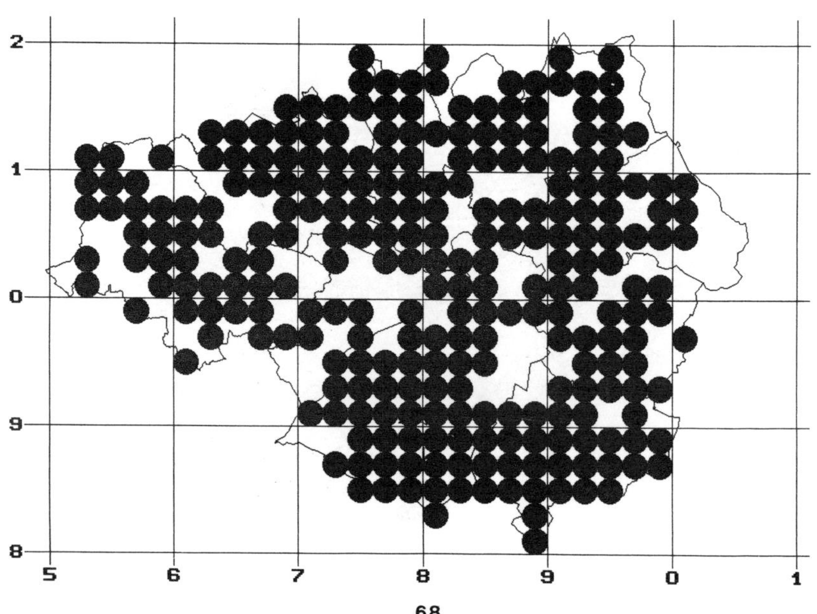

Fig.18. Callophrys rubi

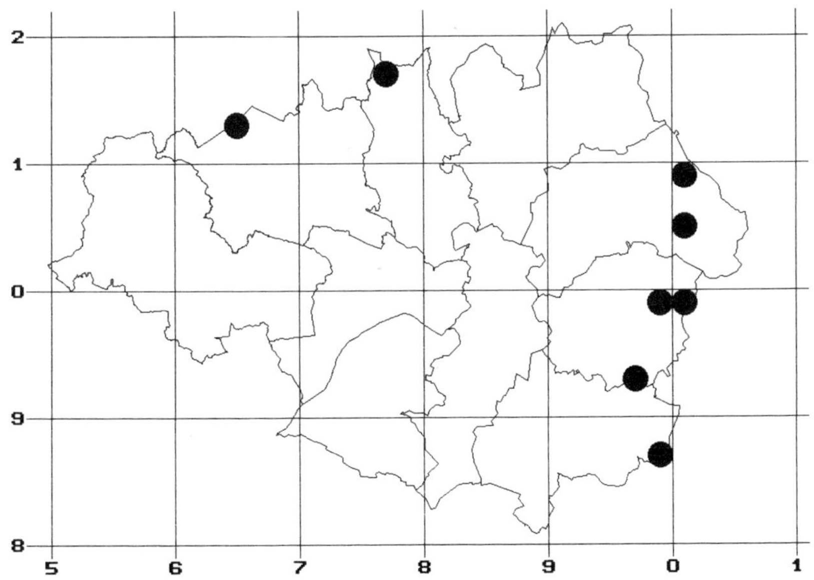

Fig.19. Quercusia quercus

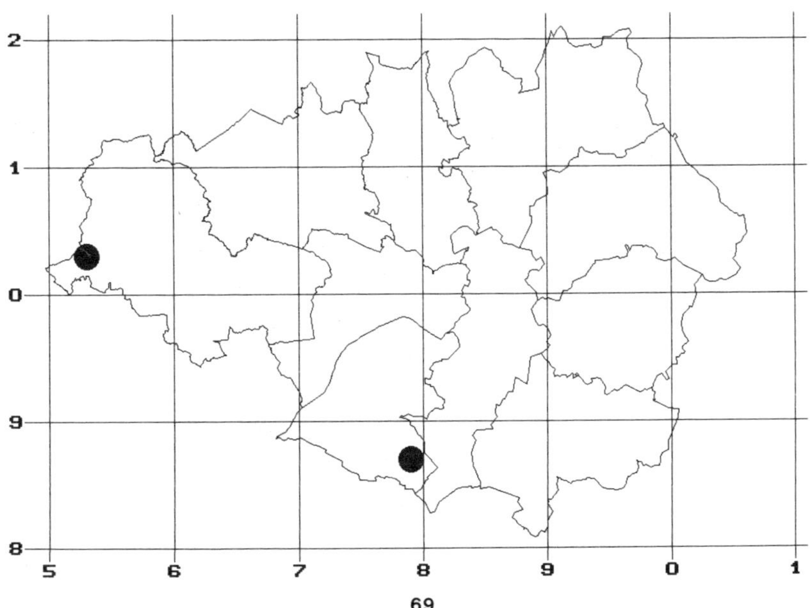

Fig.20. Satyrium w-album

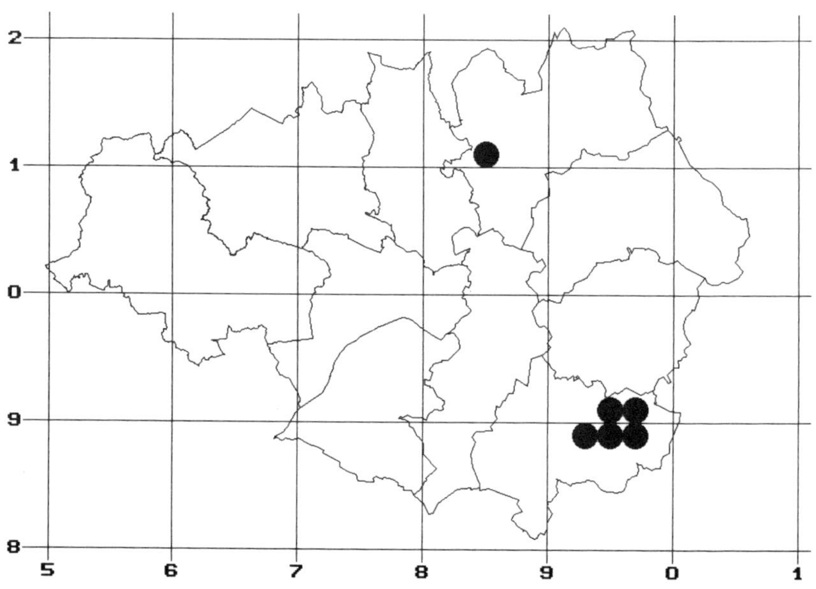

Fig.21. Lycaena phlaeas

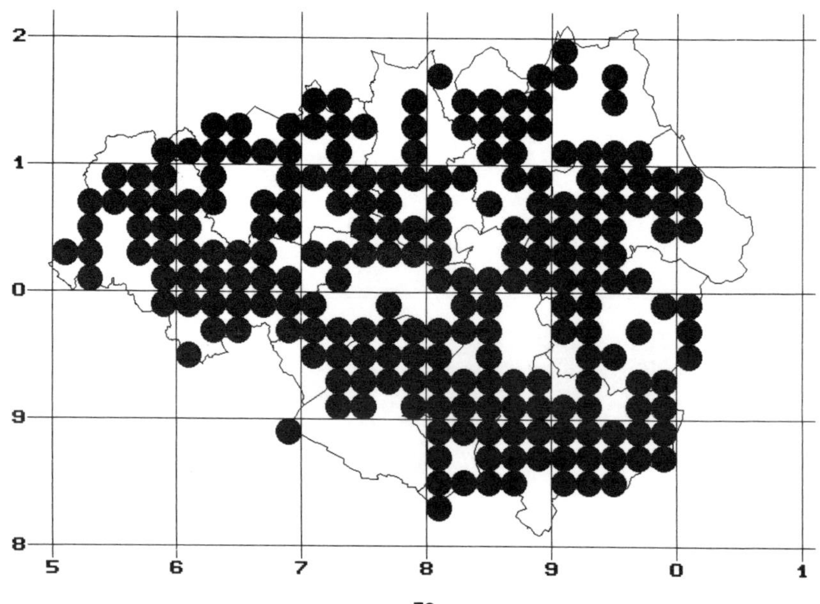

BUTTERFLY DISTRIBUTIONS, 1980-1997 - WHOLE AREA

Fig.22. Lampides boeticus

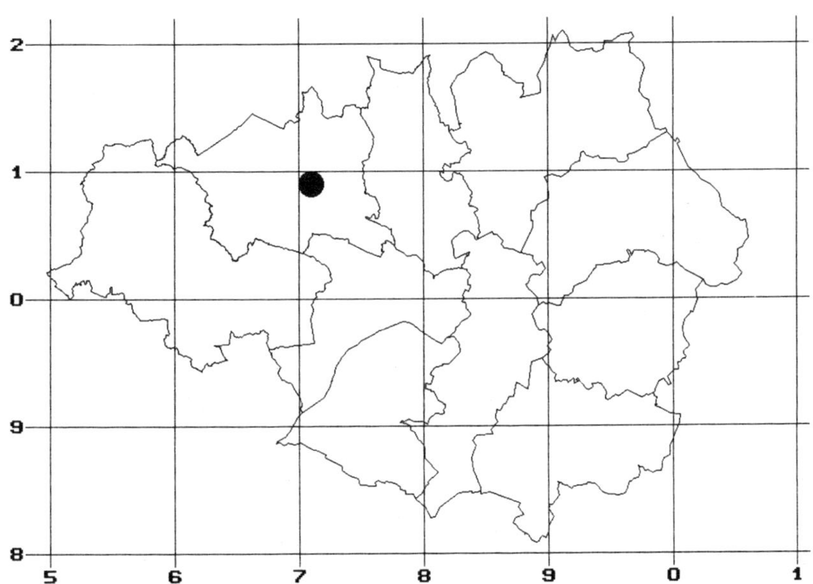

Fig.23. Polyommatus icarus

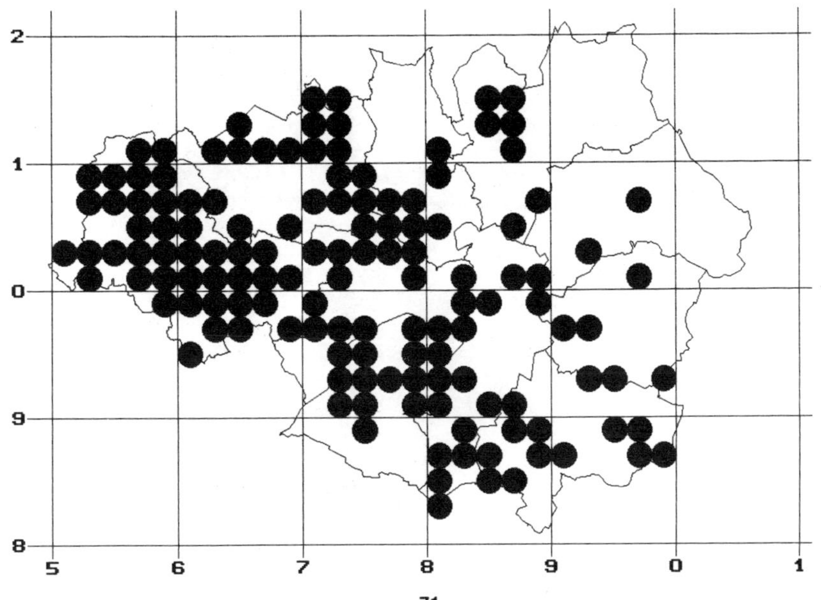

BUTTERFLY DISTRIBUTIONS, 1980-1997 - WHOLE AREA

Fig.24. Celastrina argiolus

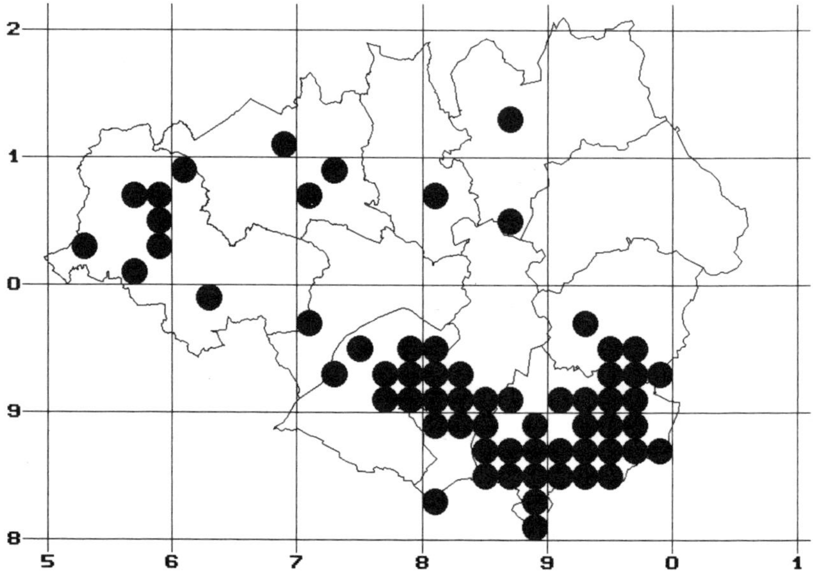

2
1
0
9
8
5 6 7 8 9 0 1

Fig.25. Vanessa atalanta

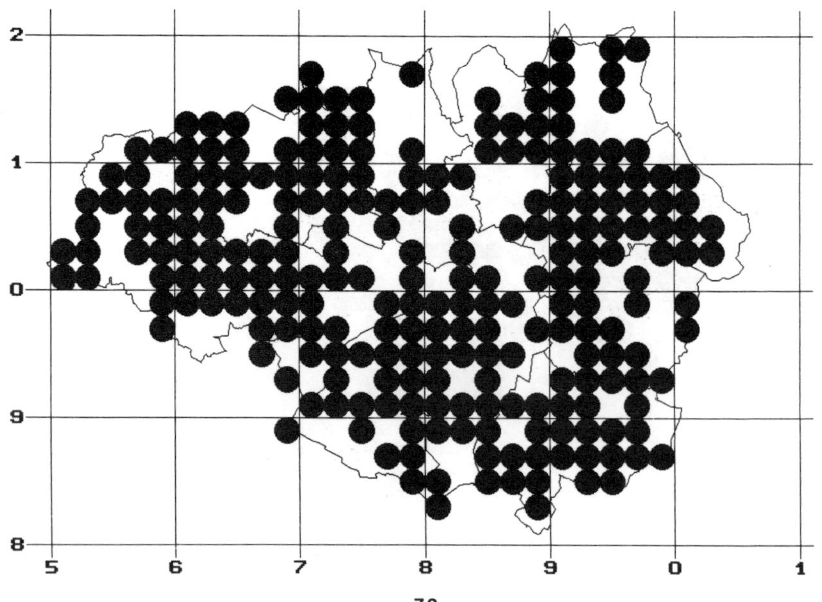

Fig.26. Vanessa cardui

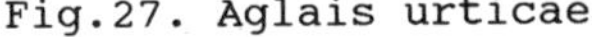

Fig.27. Aglais urticae

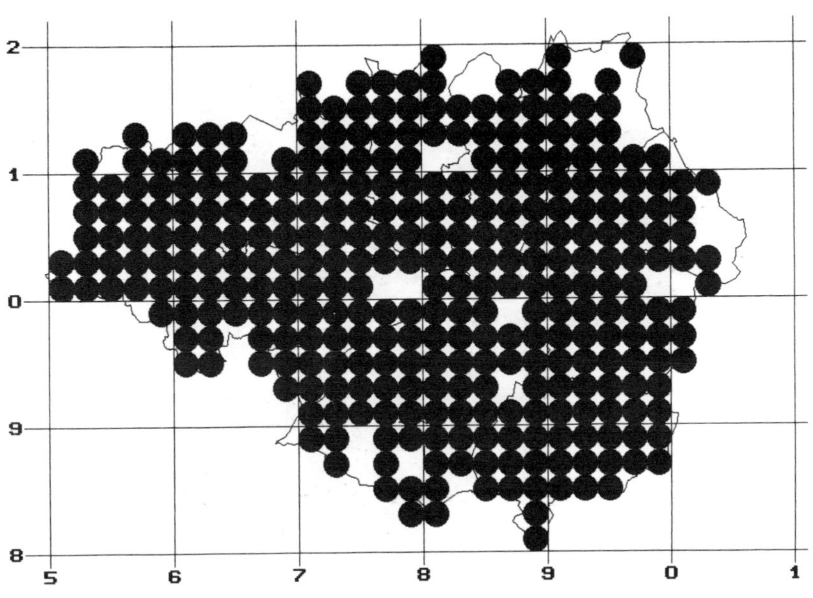

Fig.28. Nymphalis antiopa

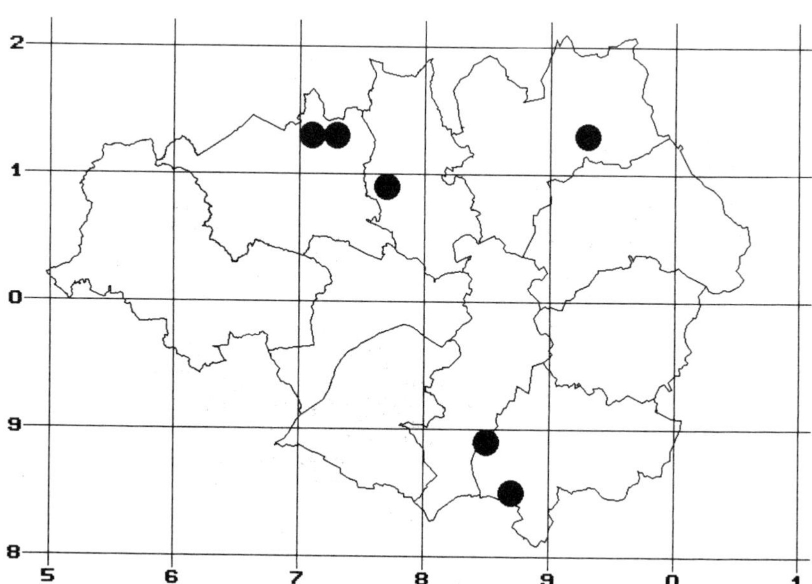

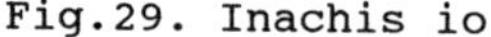

Fig.29. Inachis io

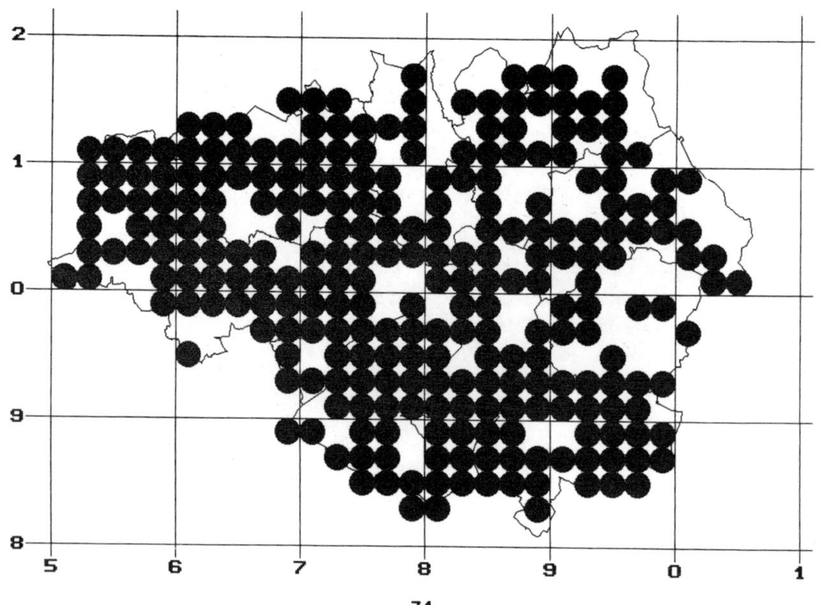

BUTTERFLY DISTRIBUTIONS, 1980-1997 - WHOLE AREA

Fig.30. Polygonia c-album

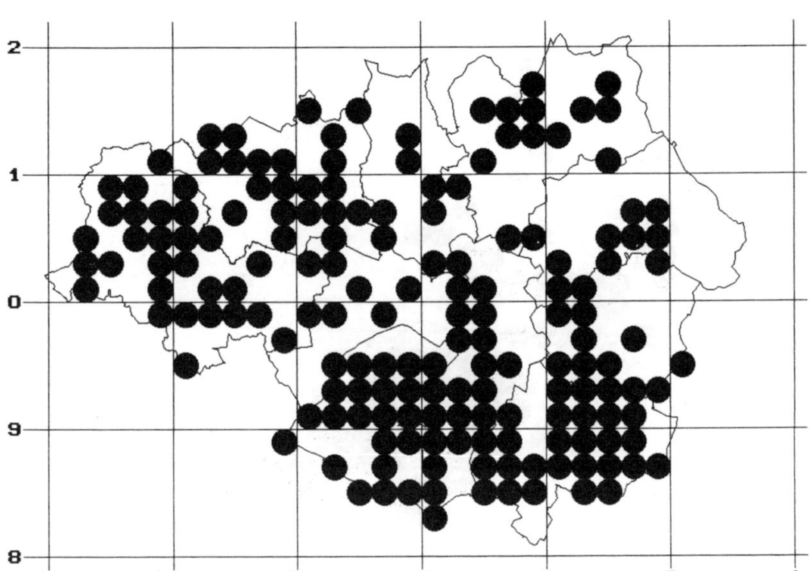

Fig.31. Argynnis aglaja

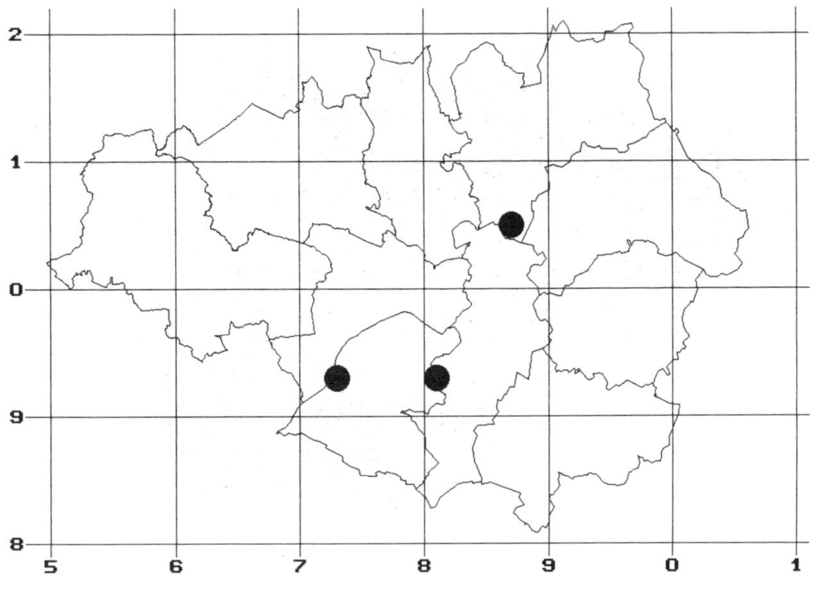

Fig.32. Pararge aegeria

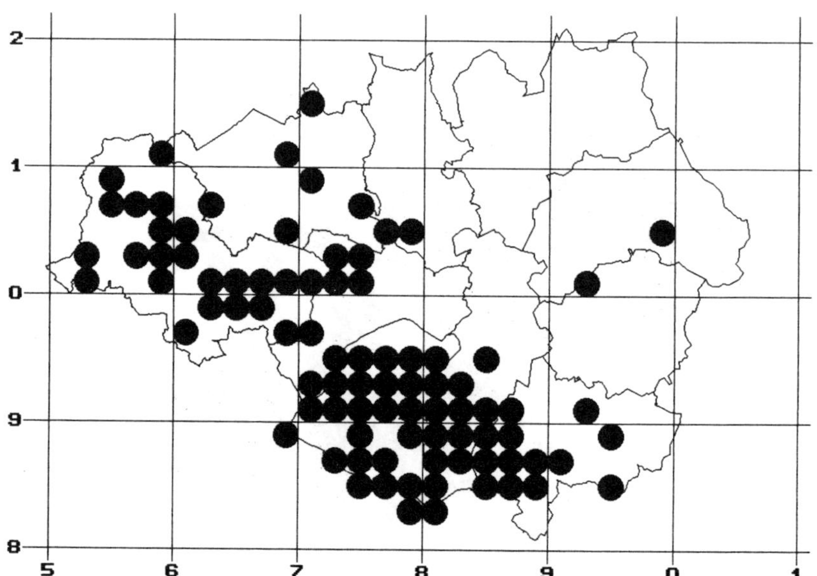

Fig.33. Lasiommata megera

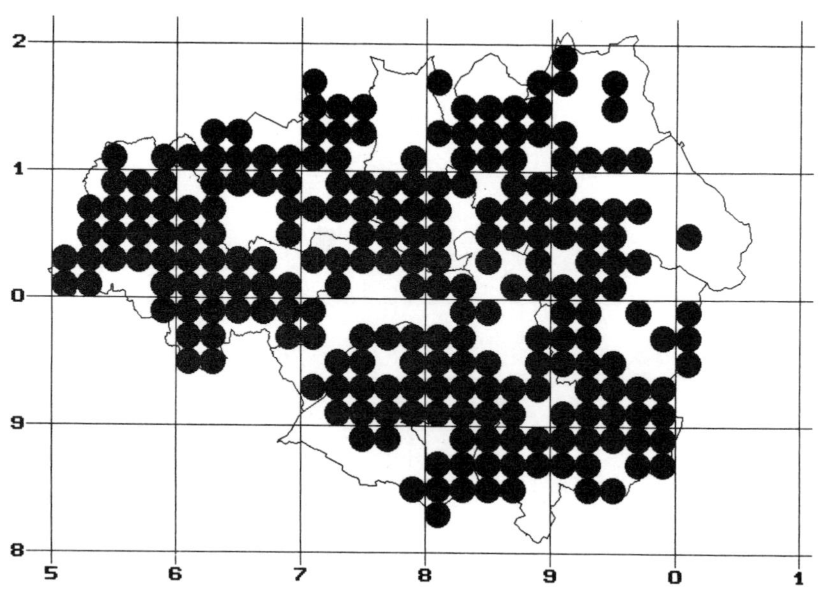

BUTTERFLY DISTRIBUTIONS, 1980-1997 - WHOLE AREA

Fig.34. Pyronia tithonus

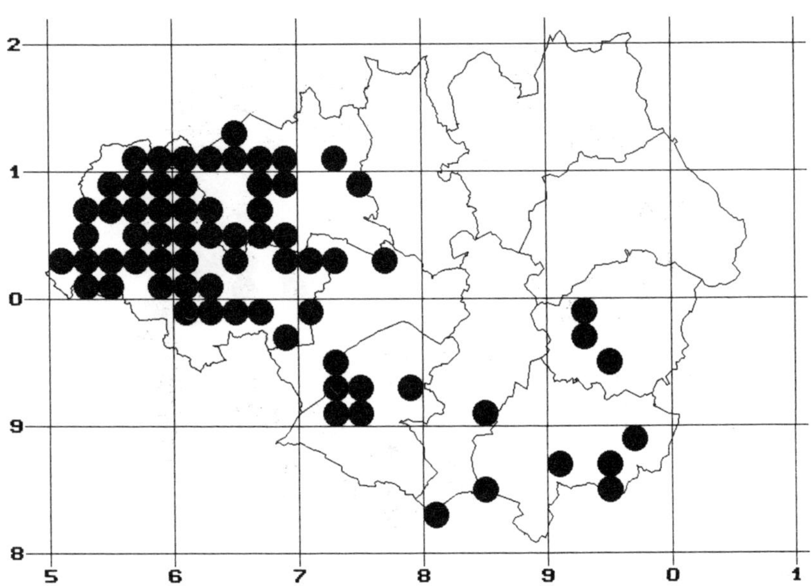

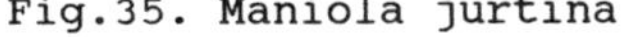
Fig.35. Maniola jurtina

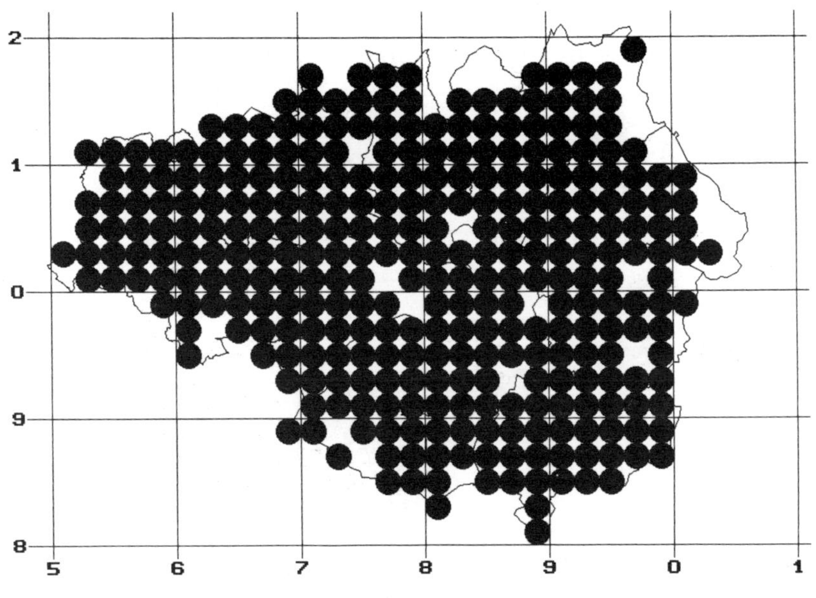

BUTTERFLY DISTRIBUTIONS, 1980-1997 - WHOLE AREA

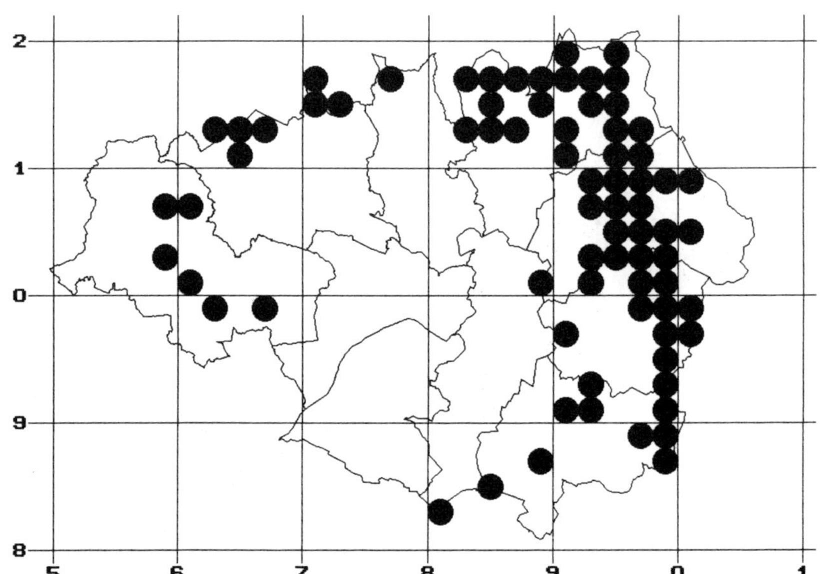

BUTTERFLY DISTRIBUTIONS - HISTORIC RECORDS

Symbols for figures 37 to 72:

- records prior to 1900 only

Open circles - records up to 1939 only

Closed circles - records up to 1979

Half-size symbols - records from 10 km square, but uncertain whether locality is inside or outside the present Greater Manchester boundary.

BUTTERFLY DISTRIBUTIONS - HISTORIC RECORDS

For explanation of symbols see page 78

Fig.37. Thymelicus sylvestris

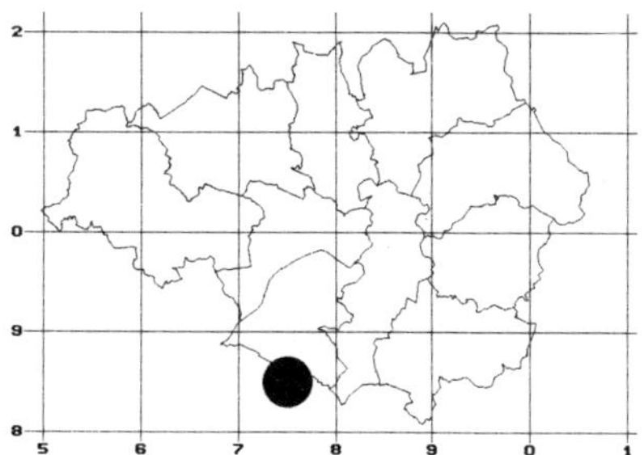

Fig.38. Ochlodes venata

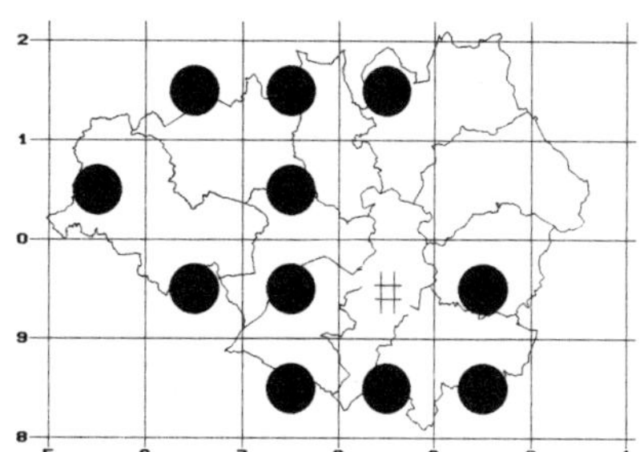

Fig.39. Erynnis tages

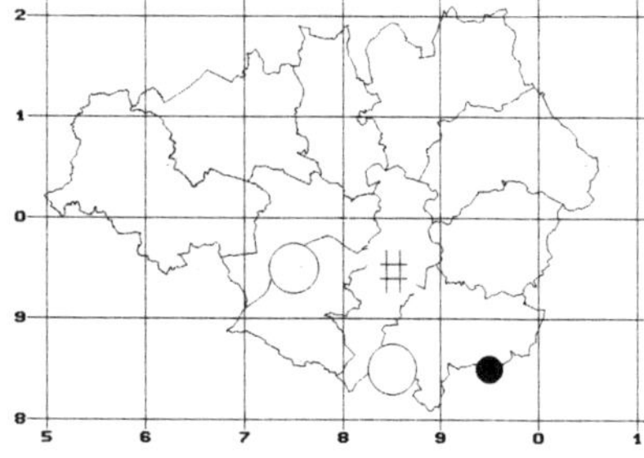

Fig.40. Leptidea sinapis

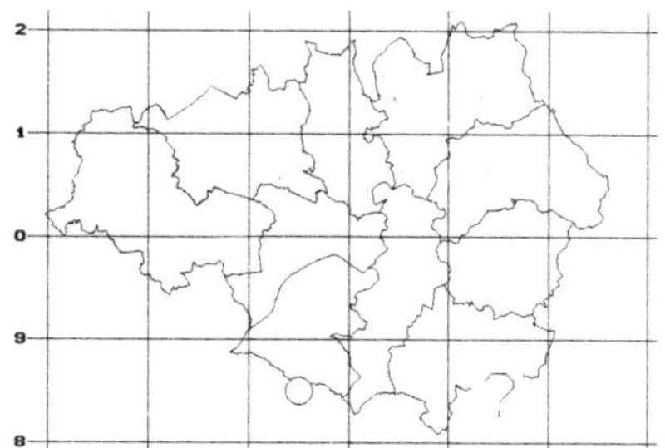

Fig.41. Colias croceus

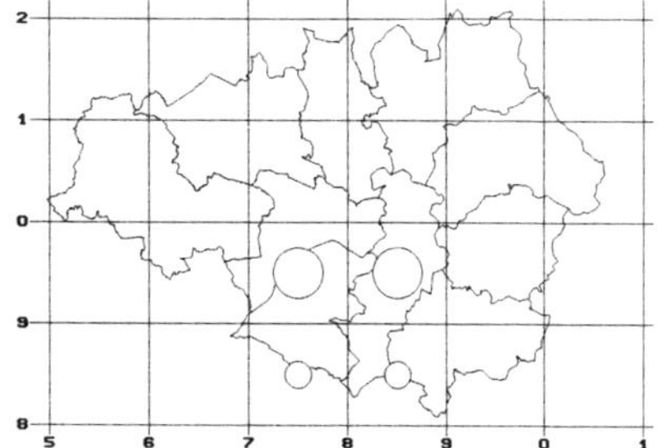

Fig.42. Gonepteryx rhamni

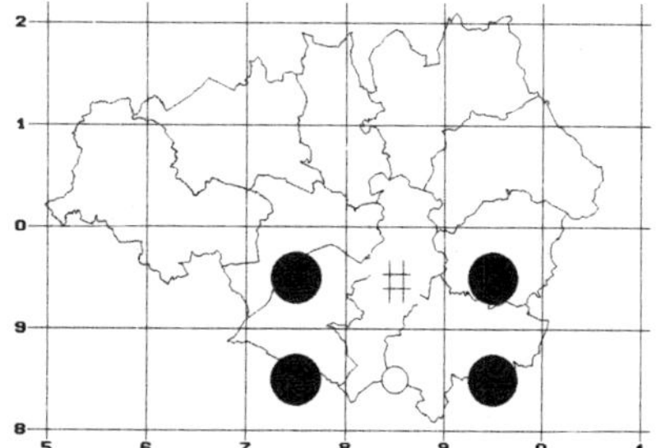

BUTTERFLY DISTRIBUTIONS - HISTORIC RECORDS

For explanation of symbols see page 78

Fig.43. Pieris brassicae

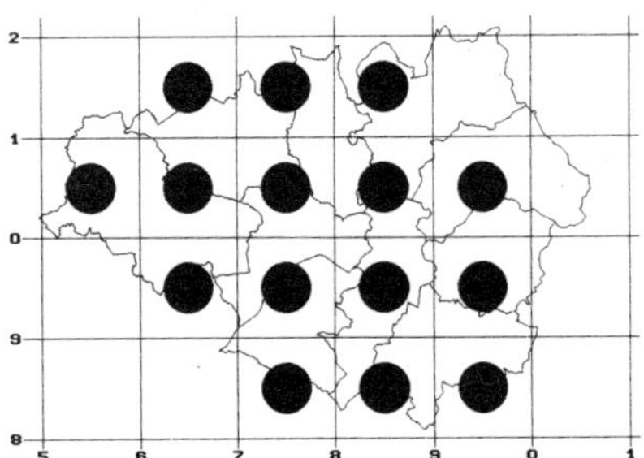

Fig.44. Pieris rapae

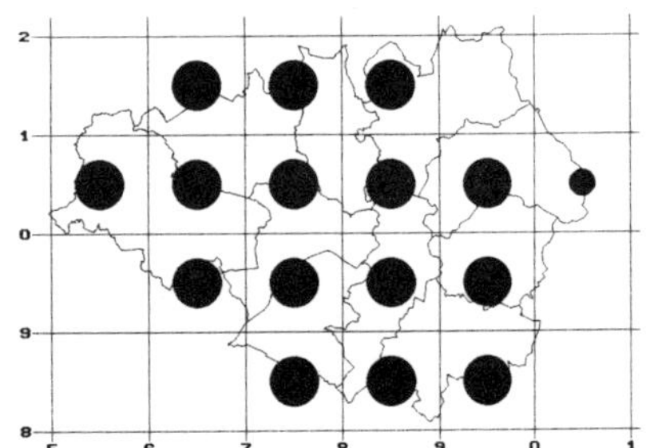

Fig.45. Pieris napi

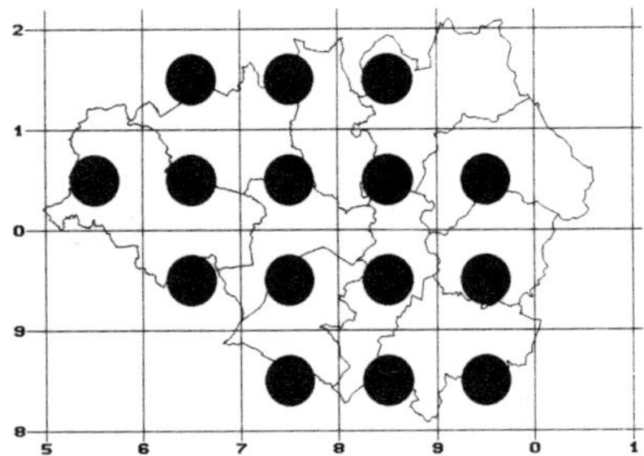

Fig.46. Anthocharis cardamines

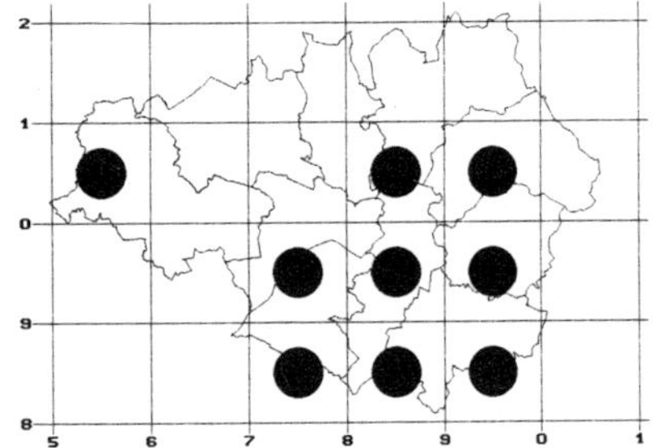

Fig.47. Callophrys rubi

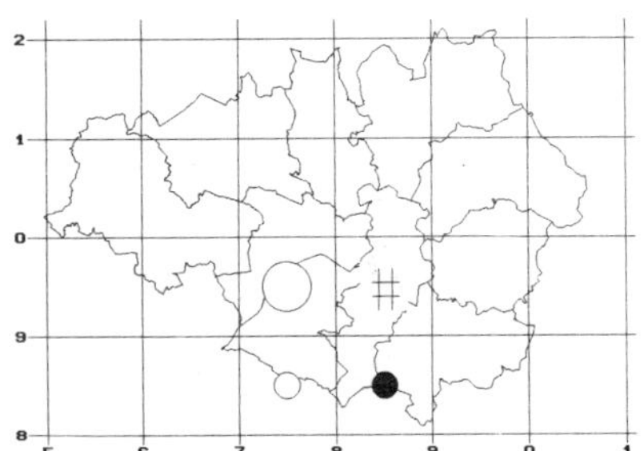

Fig.48. Quercusia quercus

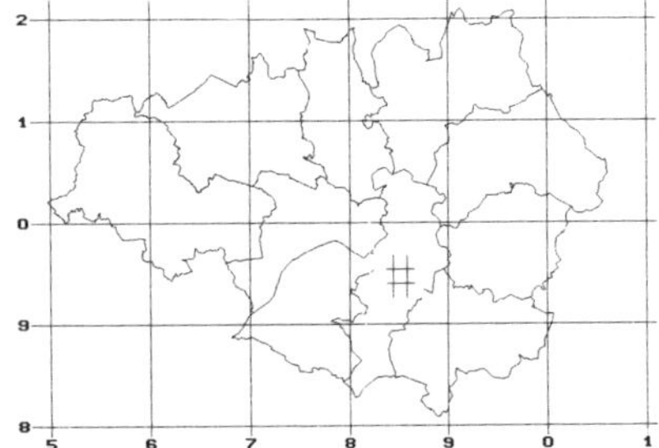

BUTTERFLY DISTRIBUTIONS - HISTORIC RECORDS

For explanation of symbols see page 78

Fig.49. Satyrium w-album

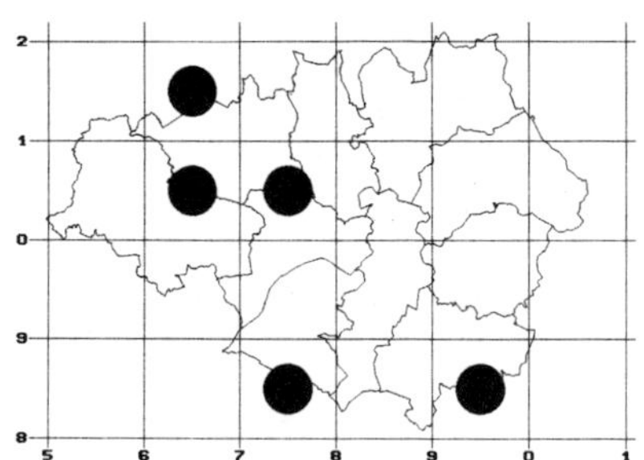

Fig.50. Lycaena phlaeas

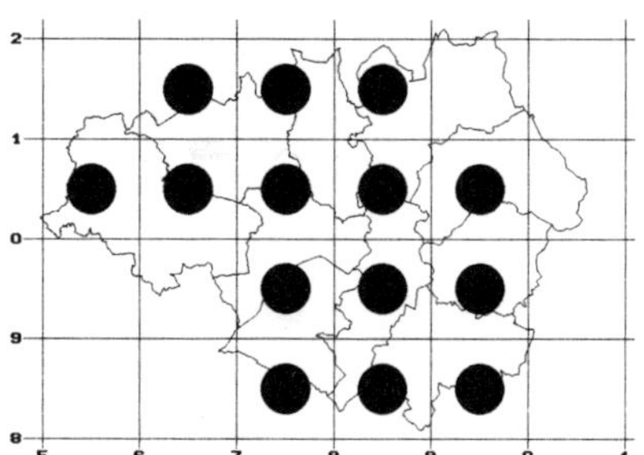

Fig.51. Cupido minimus

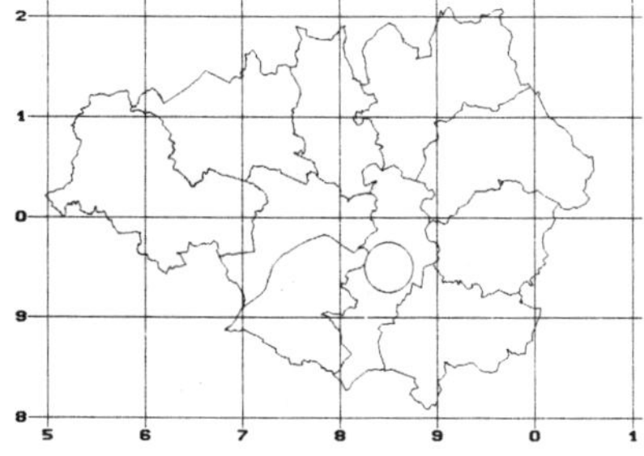

Fig.52. Plebejus argus

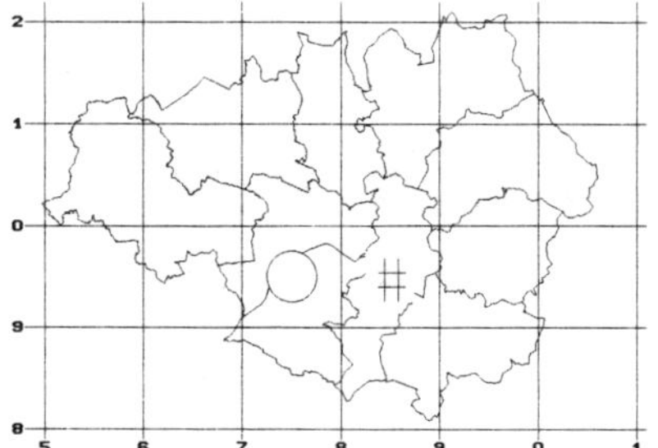

Fig.53. Aricia artaxerxes

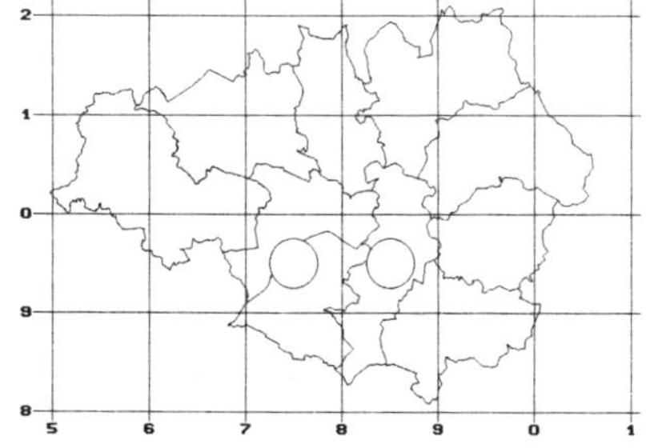

Fig.54. Polyommatus icarus

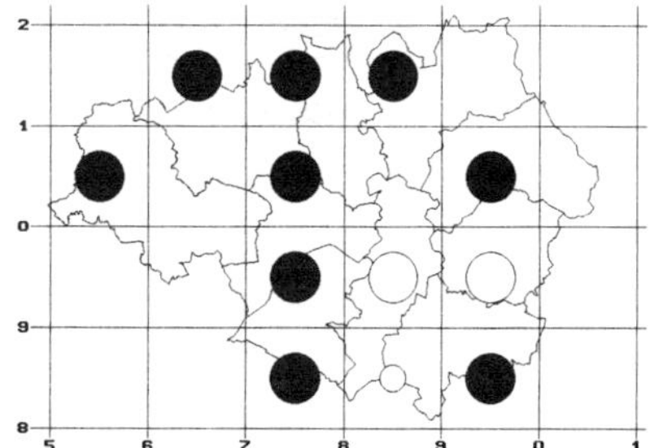

BUTTERFLY DISTRIBUTIONS - HISTORIC RECORDS

For explanation of symbols see page 78

Fig.55. Celastrina argiolus

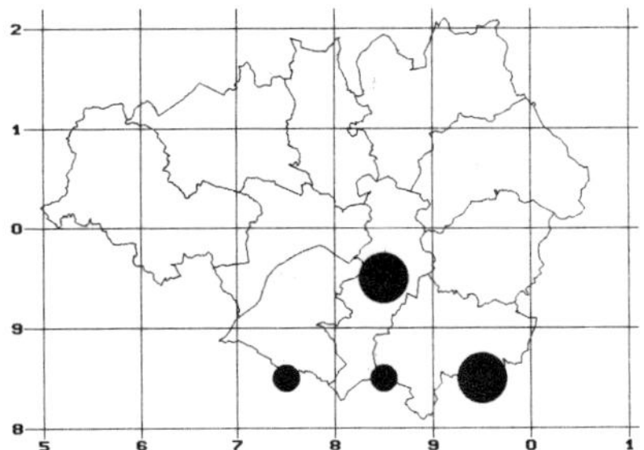

Fig.56. Vanessa atalanta

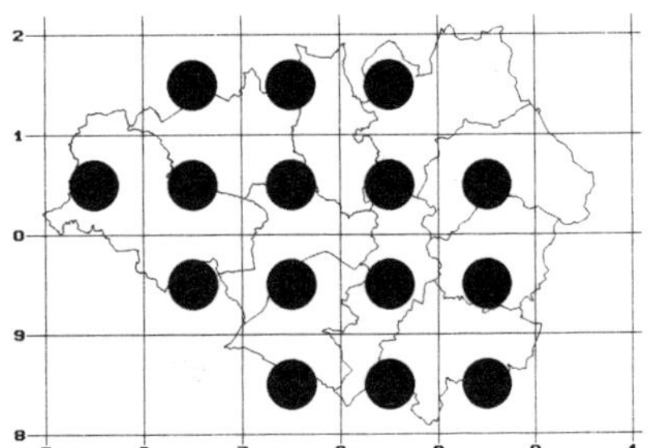

Fig.57. Vanessa cardui

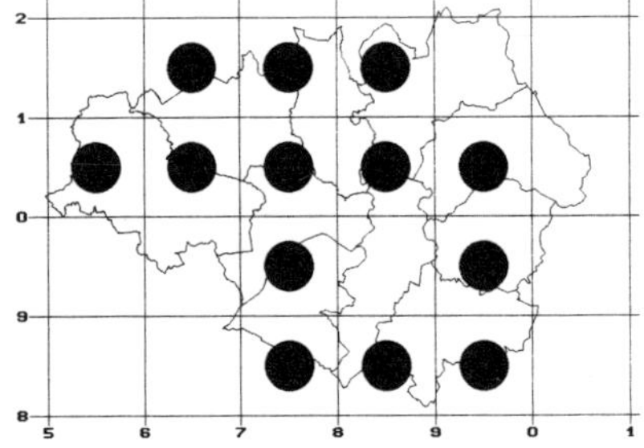

Fig.58. Aglais urticae

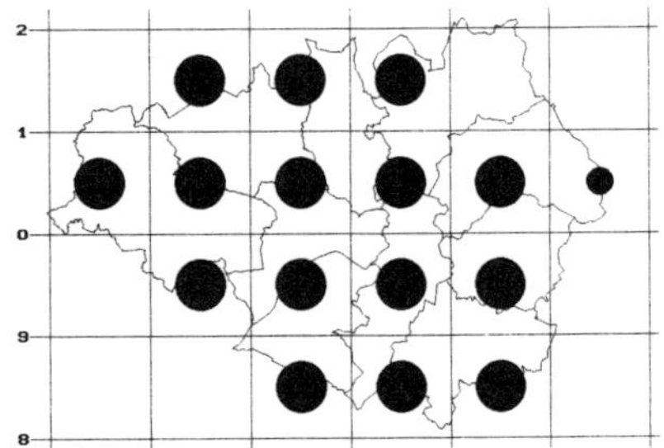

Fig.59. Nymphalis polychloros

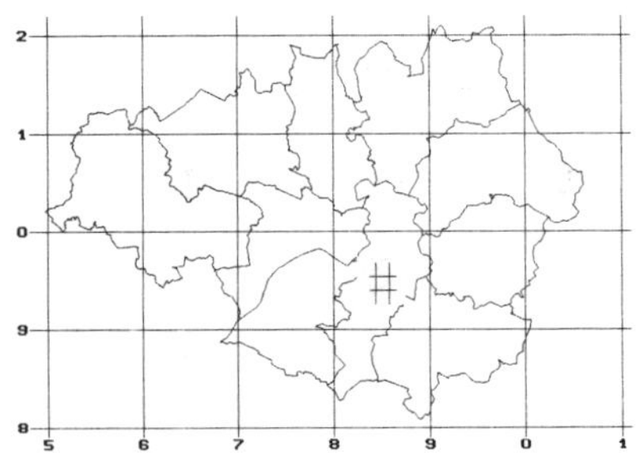

Fig.60. Inachis io

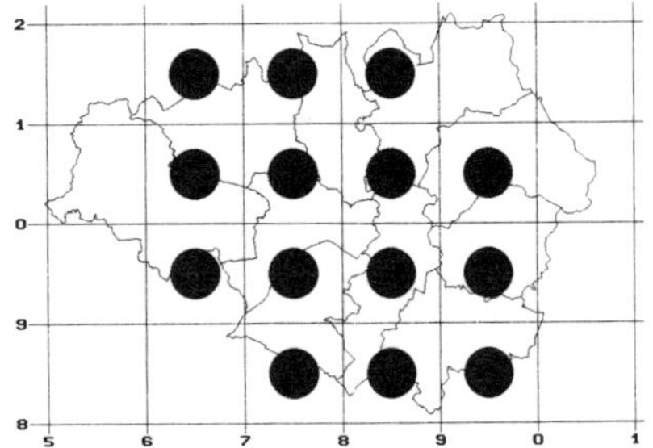

BUTTERFLY DISTRIBUTIONS - HISTORIC RECORDS

For explanation of symbols see page 78

Fig.61. Polygonia c-album

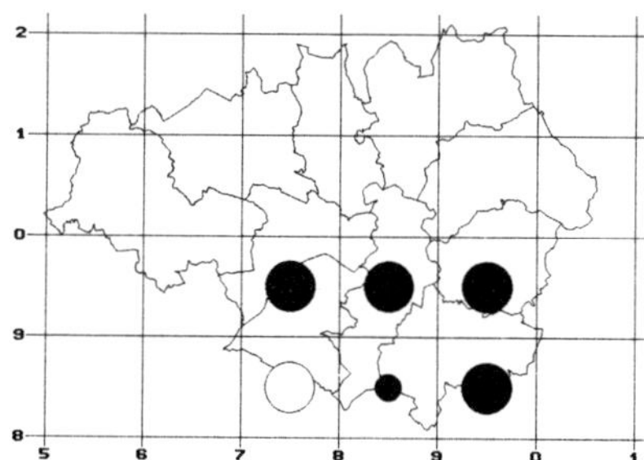

Fig.62. Boloria selene

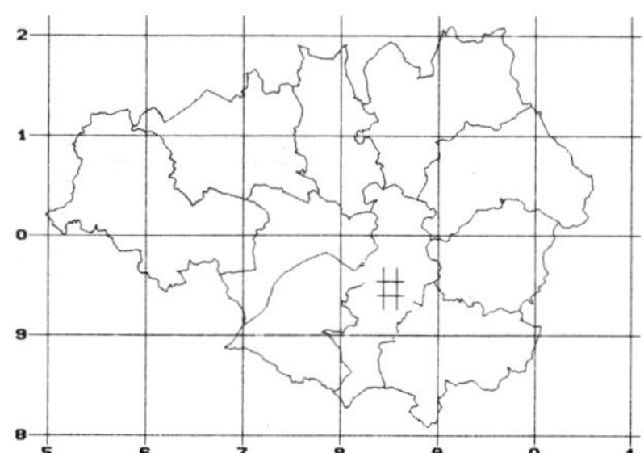

Fig.63. Boloria euphrosyne

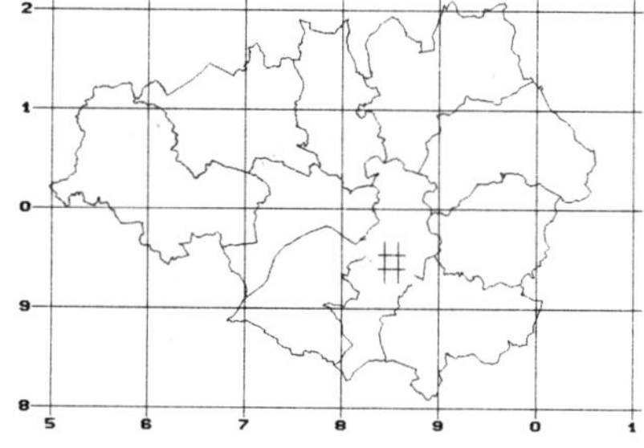

Fig.64. Argynnis paphia

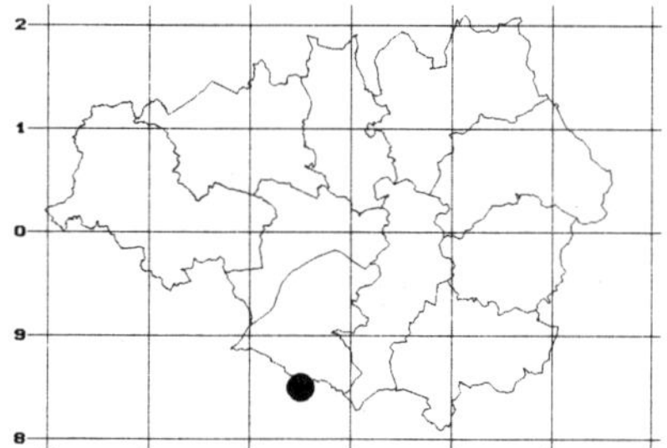

Fig.65. Euphydryas aurinia

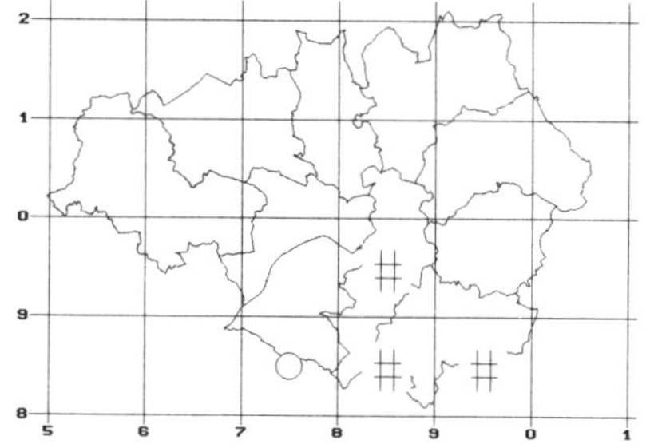

Fig.66. Pararge aegeria

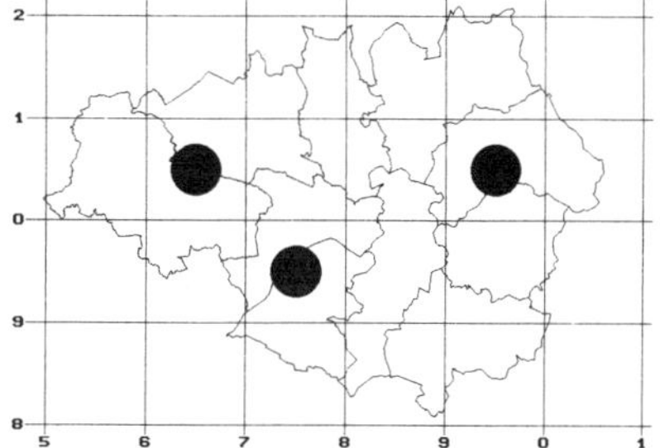

BUTTERFLY DISTRIBUTIONS - HISTORIC RECORDS

For explanation of symbols see page 78

Fig.67. Lasiommata megera

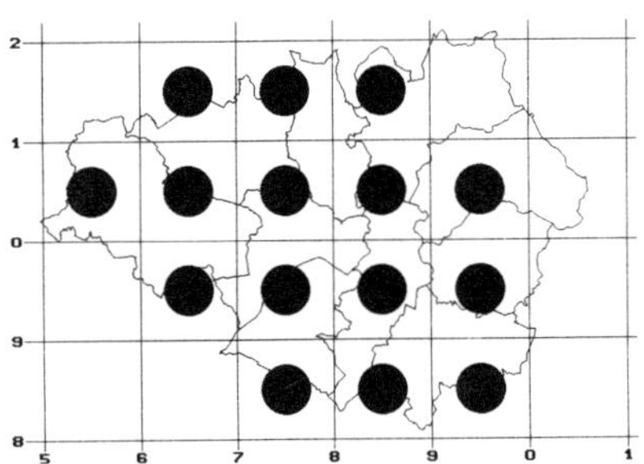

Fig.68. Hipparchia semele

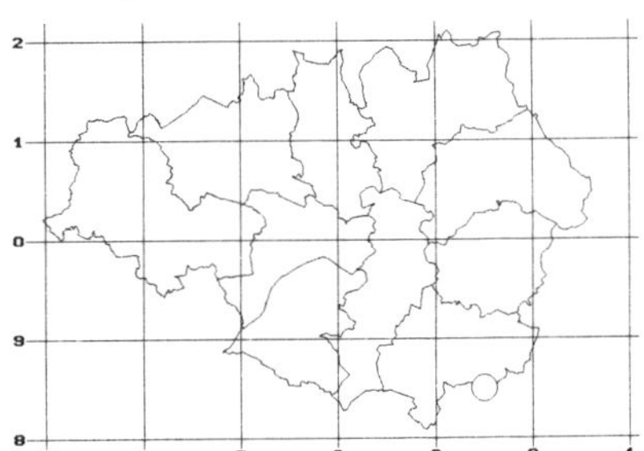

Fig.69. Pyronia tithonus

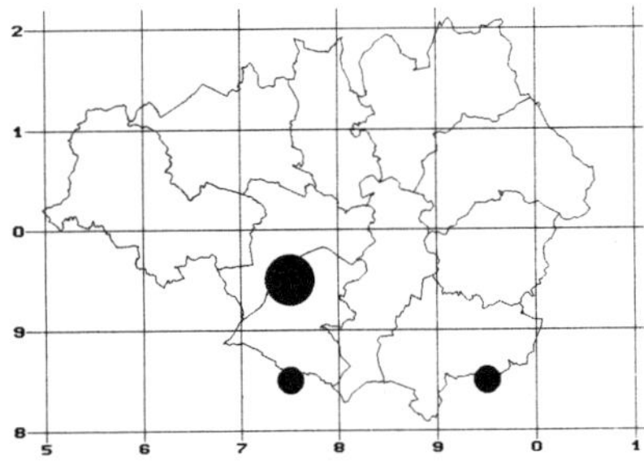

Fig.70. Maniola jurtina

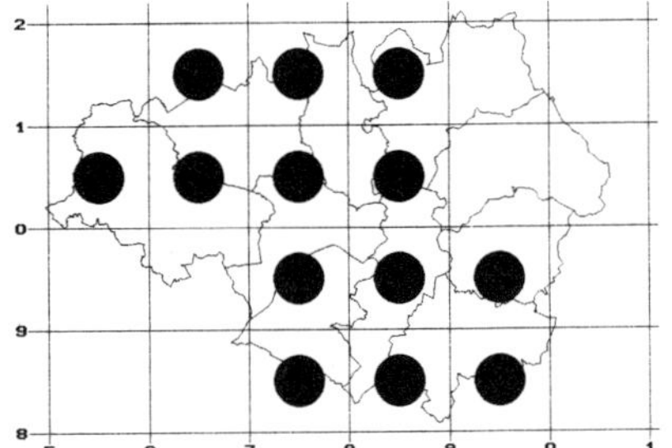

Fig.71. Coenonympha pamphilus

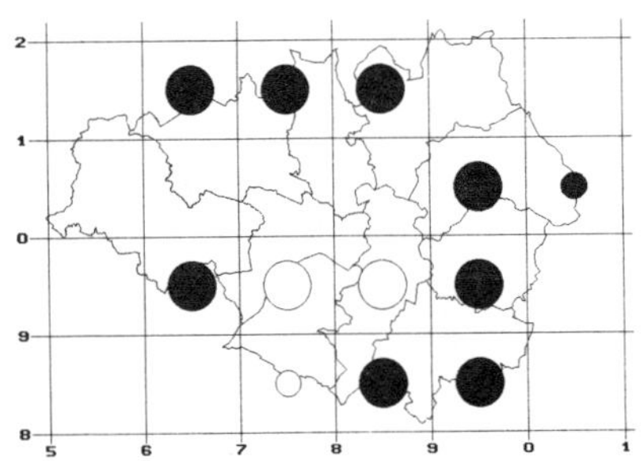

Fig.72. Coenonympha tullia

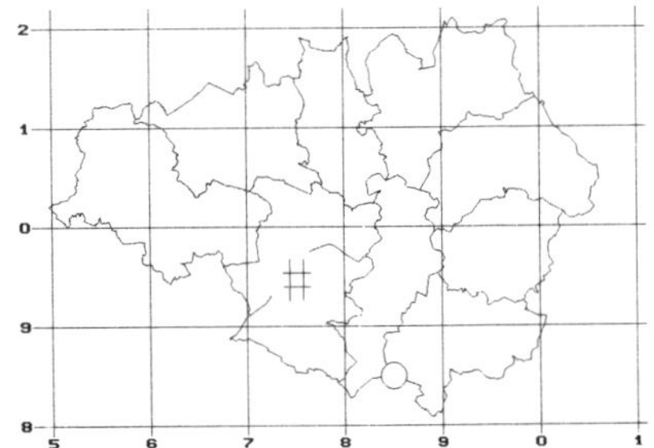

ENVIRONMENTS - 7 X 5 KM ZONE

Fig.73. THE 7 X 5 KM ZONE

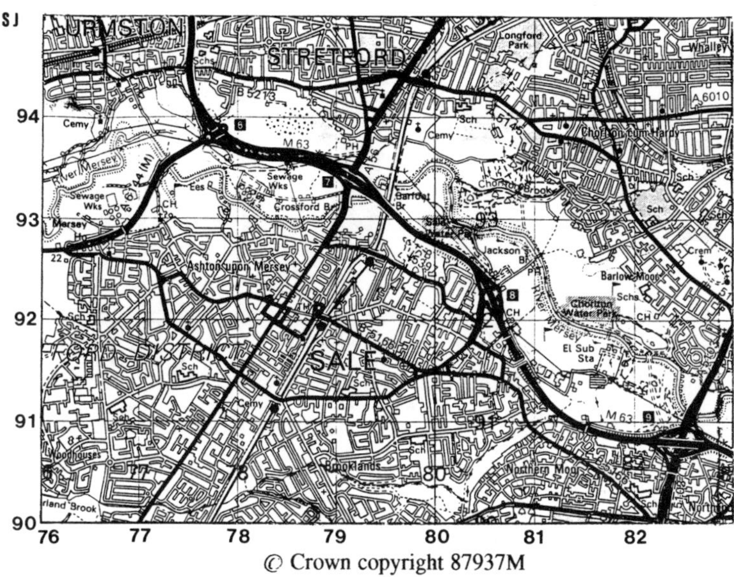

Fig.74. GRASS AND SEMI-NATURAL VEGETATION POTENTIALLY SUITABLE AS HABITAT

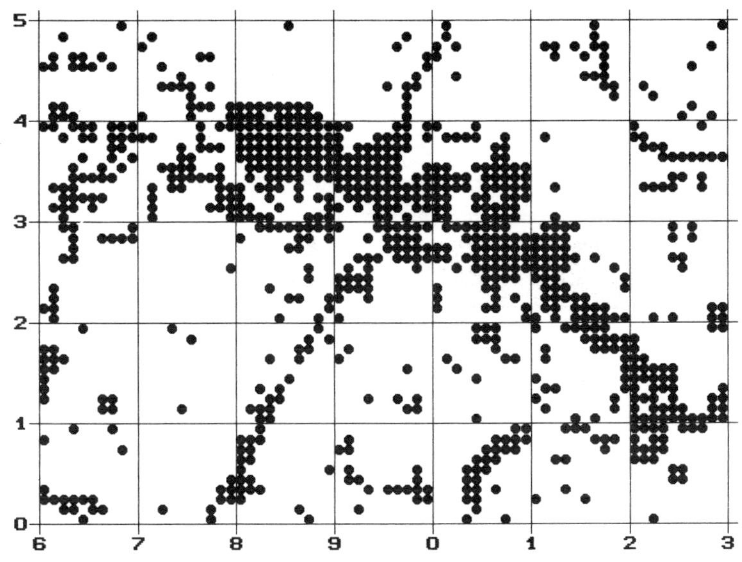

ENVIRONMENTS - 7 X 5 KM ZONE

Fig.75. HEAVILY GRAZED SITES UNSUITABLE AS HABITAT

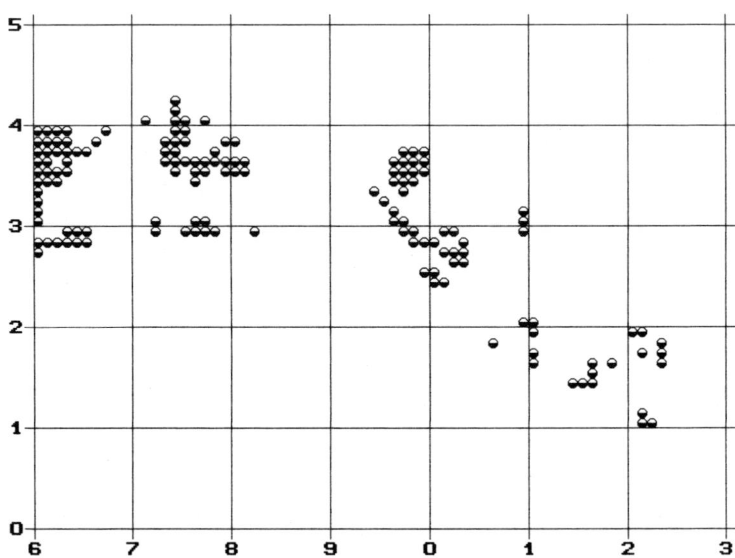

Fig.76. FORMAL OPEN SPACES WITH MOWN GRASS, UNSUITABLE AS HABITAT

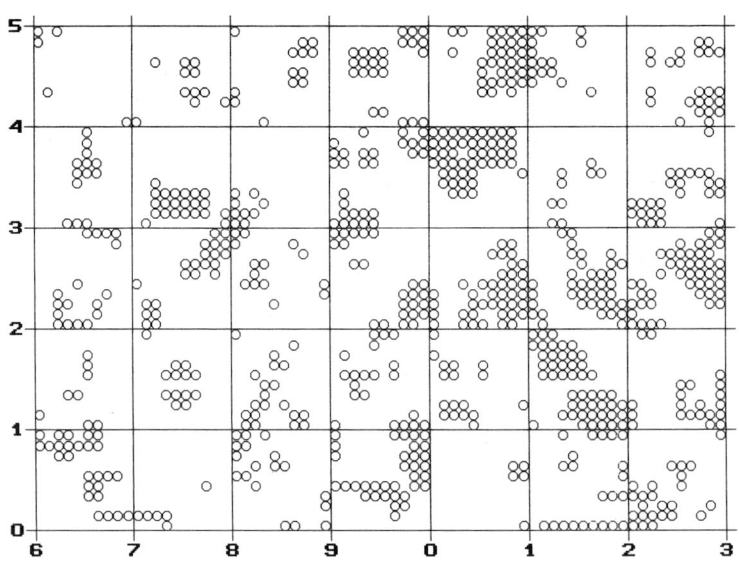

URBAN COVER - 7 X 5 KM ZONE

Full-size symbols, 50-100% cover. Half-size symbols, less than 50% cover.

Fig.77. URBAN COVER in 1996

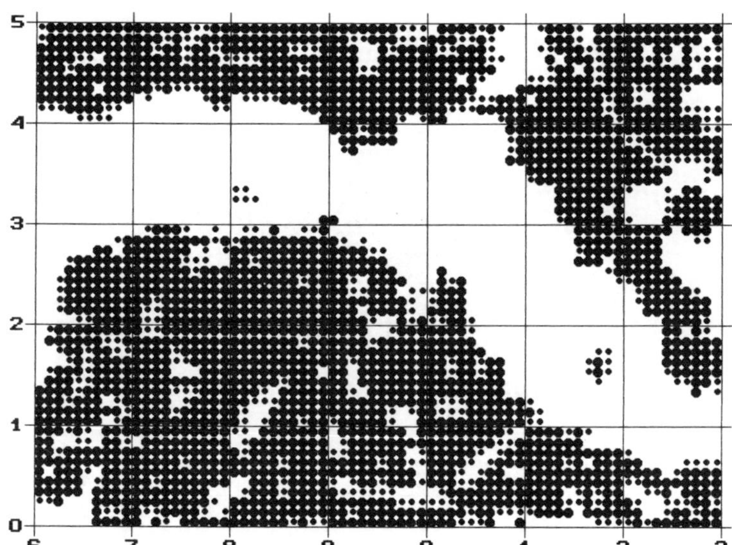

Fig.78. URBAN COVER PRIOR TO 1960

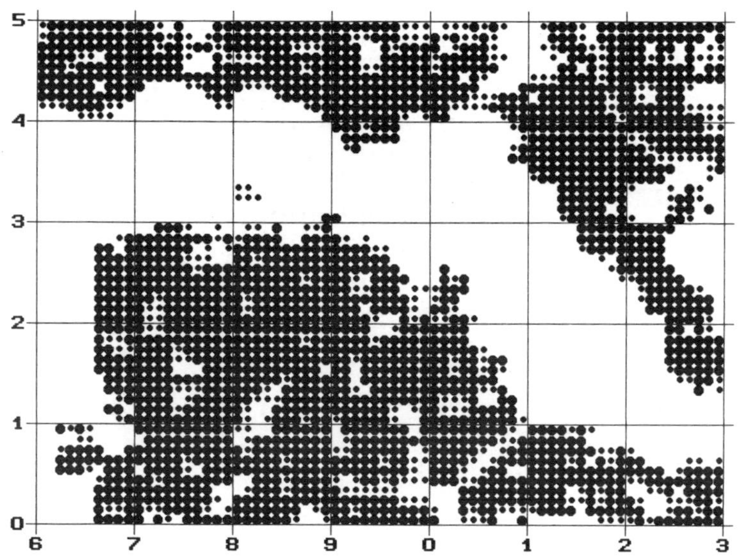

RECORDING COVERAGE - 7 X 5 KM ZONE

Open circles in fig.79, visits in winter only. Other symbols, graded from smallest size (dot) 1 visit during period to largest size 100+ visits.

Fig.79. RECORDING COVERAGE 1994-7, WHOLE YEAR

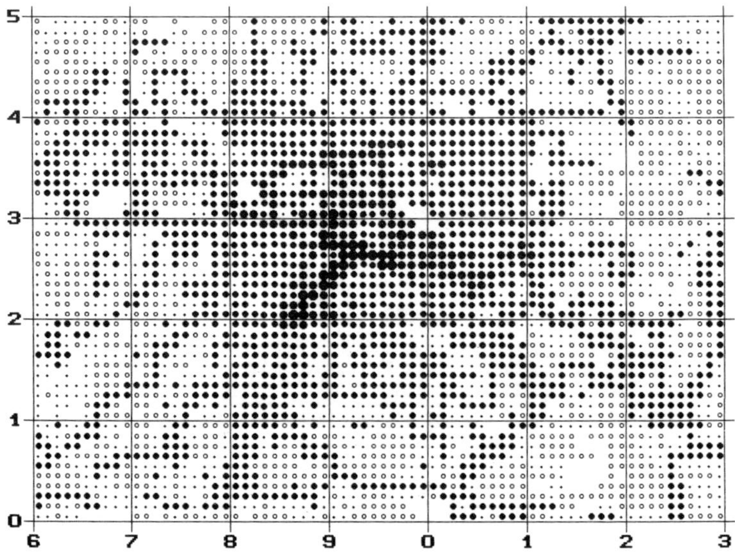

Fig.80. RECORDING COVERAGE 1994-7, DURING BUTTERLY SEASON

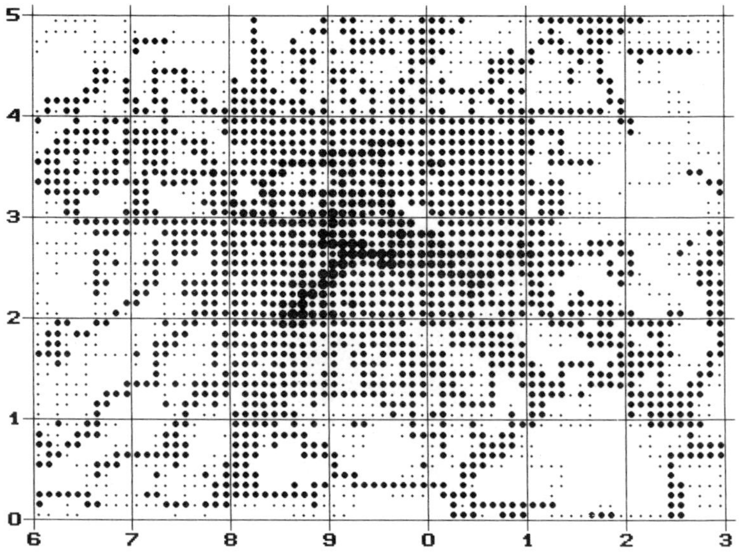

BUTTERFLY DISTRIBUTIONS, 1992-1997 - 7 X 5 KM ZONE

Fig.81. RECORDS OF ALL BUTTERFLY SPECIES IN 7 X 5 KM ZONE

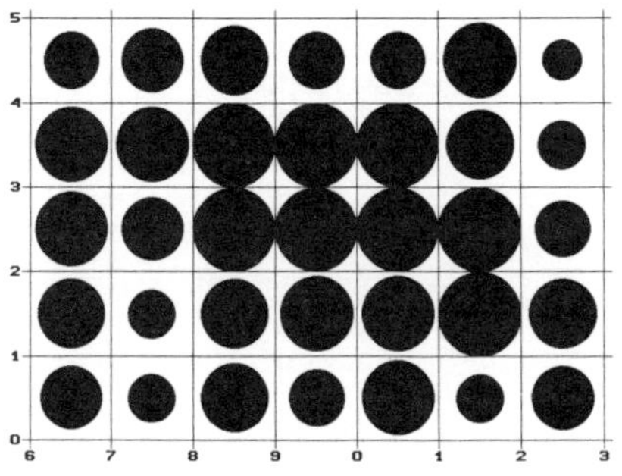

7	9	12	7	8	13	3
13	14	20	20	16	12	6
13	10	19	19	17	17	7
11	6	12	13	14	17	12
10	6	11	7	14	6	9

In the following figures 82 to 101, the symbols denote the percentage of the 100 m squares visited, within the 1 km square, in which the butterfly species was recorded:

1-5%

6-10%

11-20%

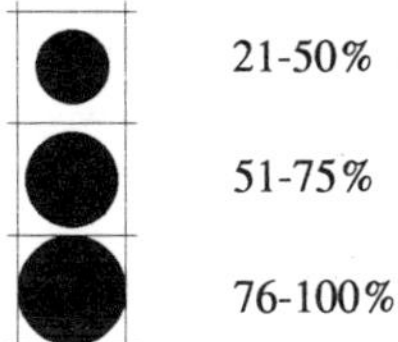

21-50%

51-75%

76-100%

Fig.82. Thymelicus sylvestris

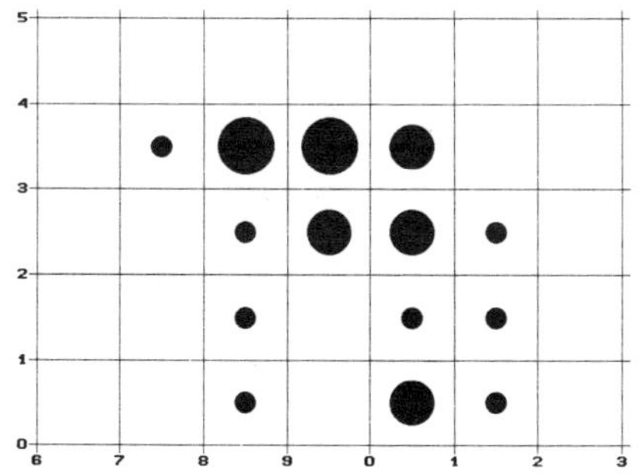

Fig.83. Ochlodes venata

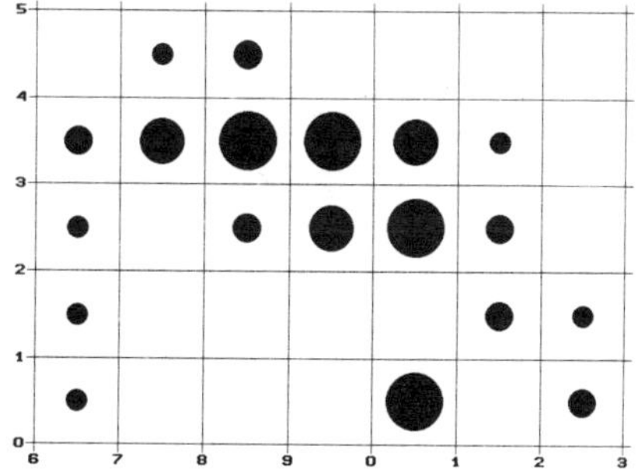

BUTTERFLY DISTRIBUTIONS, 1992-1997 - 7 X 5 KM ZONE

For explanation of symbols see page 89

Fig.84. Colias croceus

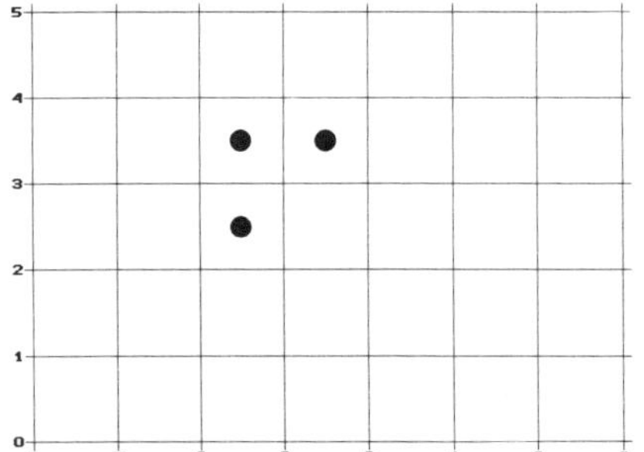

Fig.85. Gonepteryx rhamni

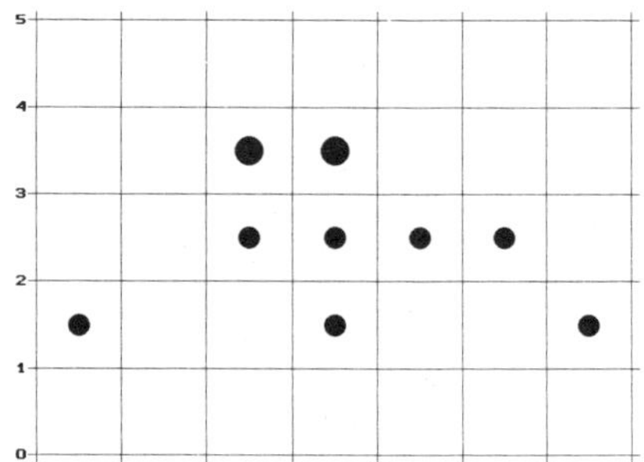

Fig.86. Pieris brassicae

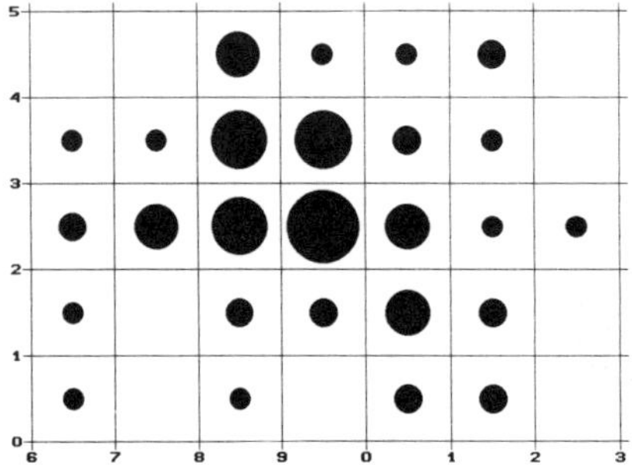

Fig.87. Pieris rapae

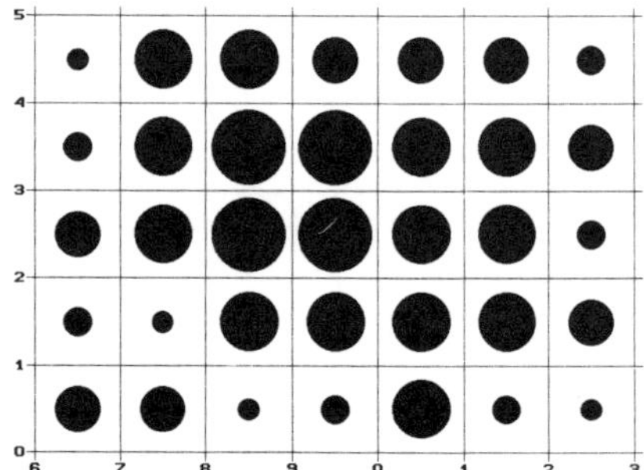

Fig.88. Pieris napi

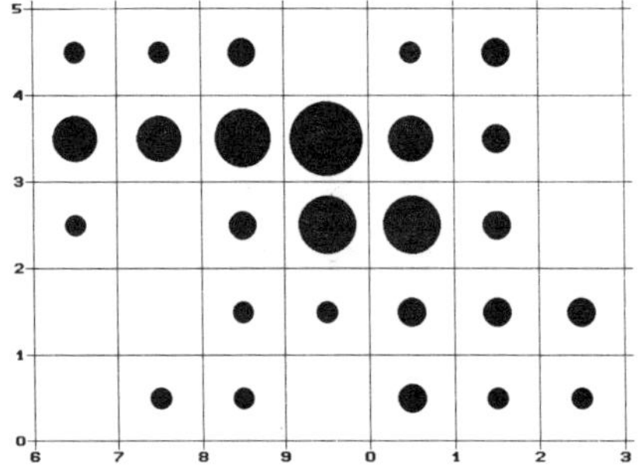

Fig.89. Anthocharis cardamines

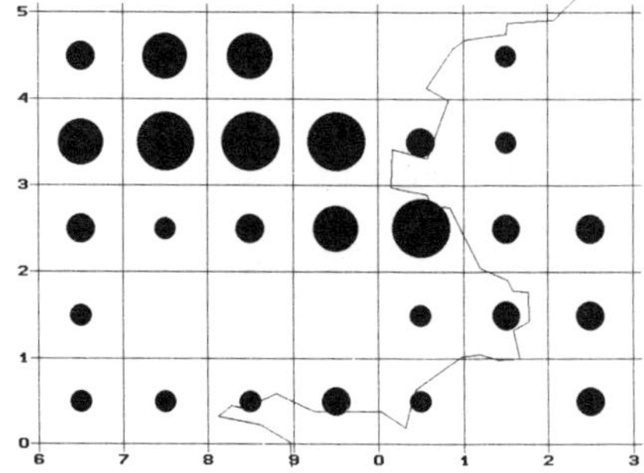

BUTTERFLY DISTRIBUTIONS, 1992-1997 - 7 X 5 KM ZONE

For explanation of symbols see page 89

Fig.90. Lycaena phlaeas

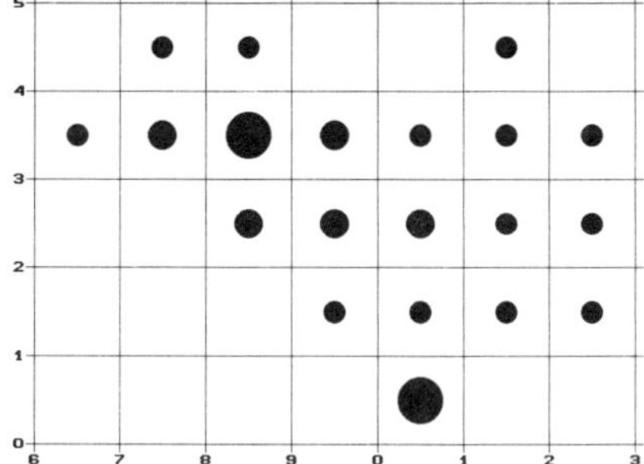

Fig.91. Polyommatus icarus

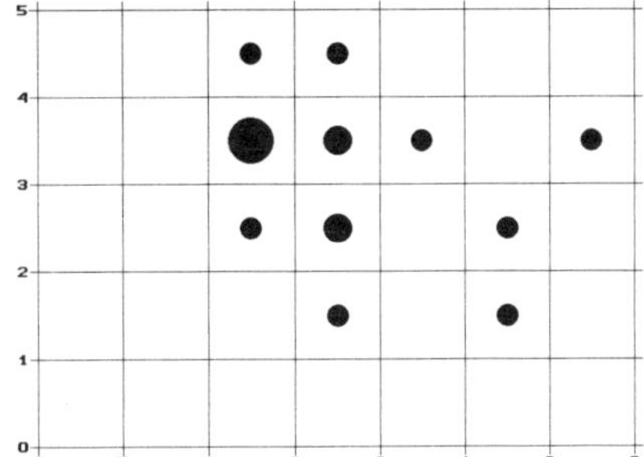

Fig.92. Celastrina argiolus

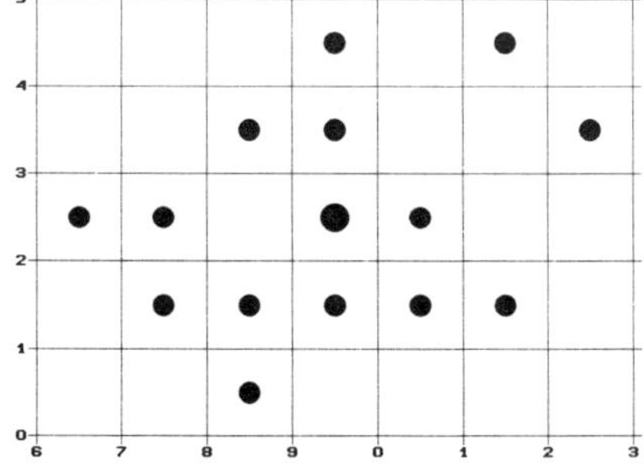

Fig.93. Vanessa atalanta

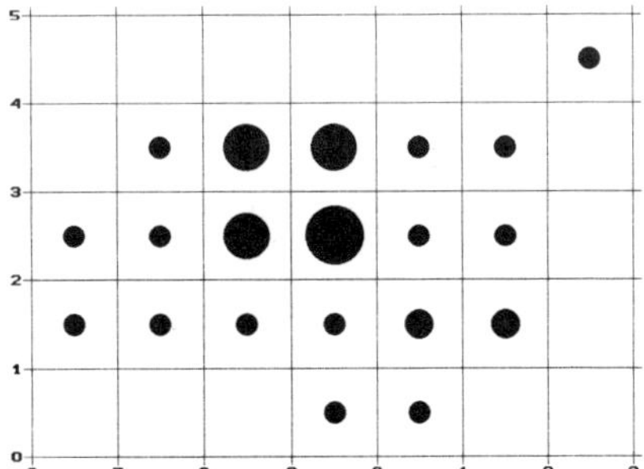

Fig.94. Vanessa cardui

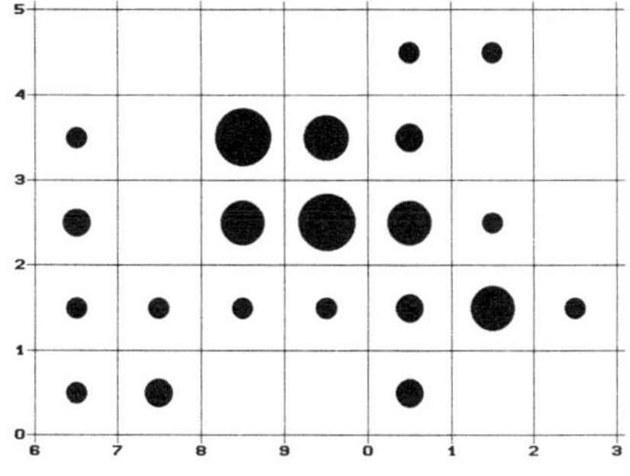

Fig.95. Aglais urticae

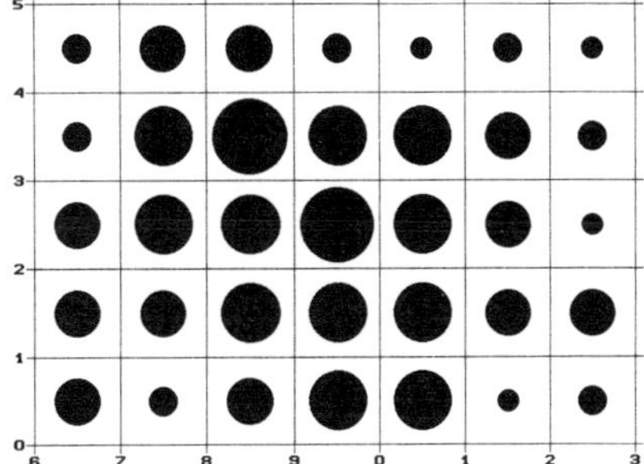

BUTTERFLY DISTRIBUTIONS, 1992-1997 - 7 X 5 KM ZONE

For explanation of symbols see page 89

Fig.96. Inachis io

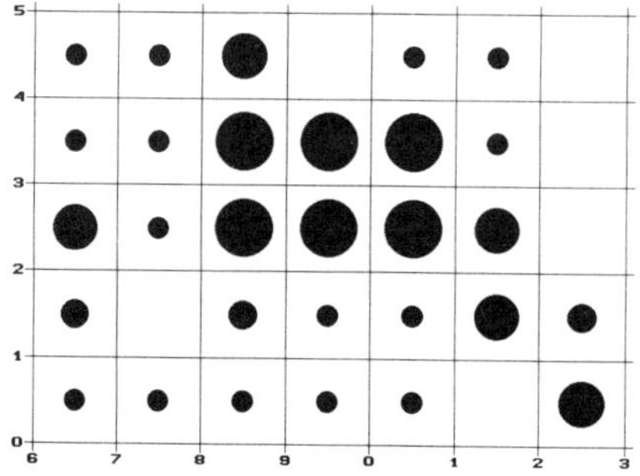

Fig.97. Polygonia c-album

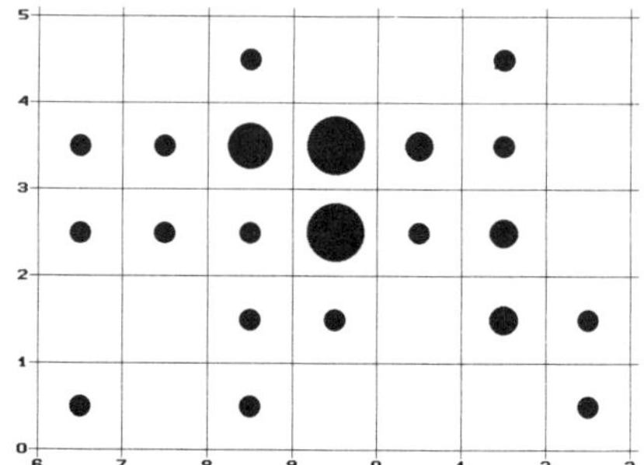

Fig.98. Pararge aegeria

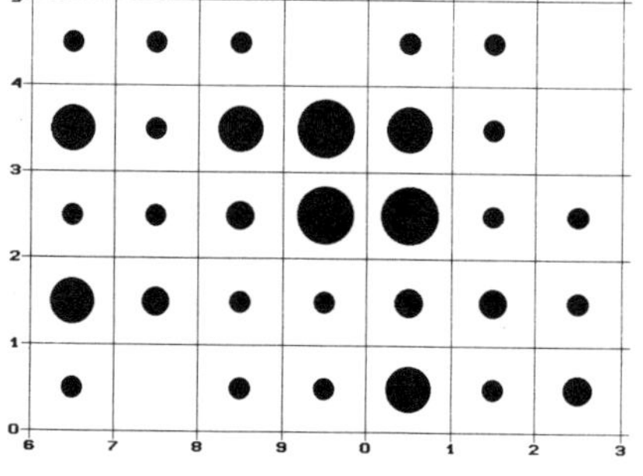

Fig.99. Lasiommata megera

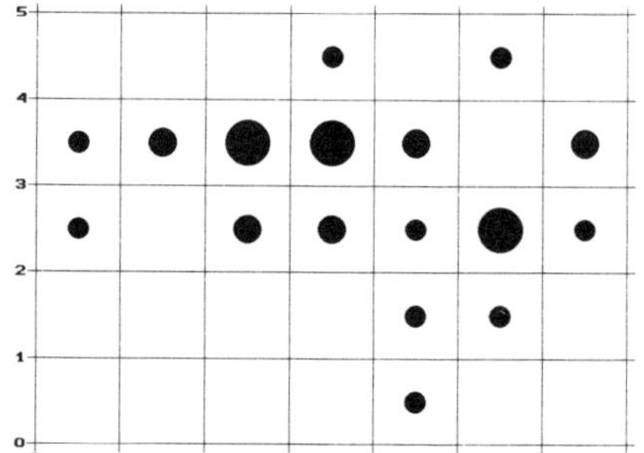

Fig.100. Pyronia tithonus

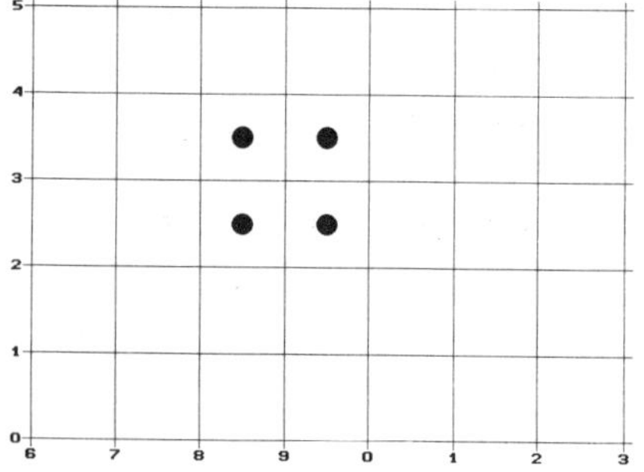

Fig.101. Maniola jurtina

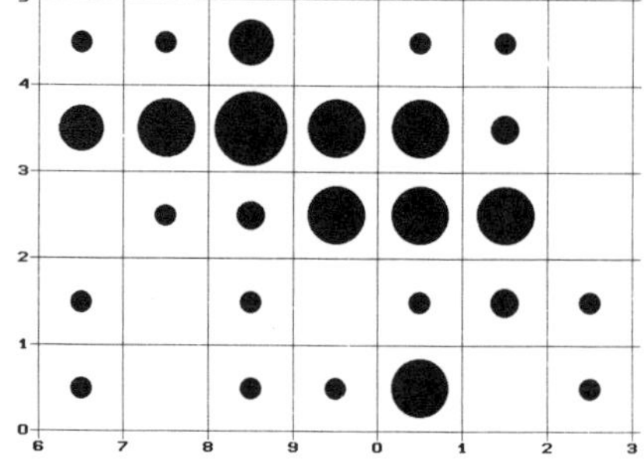

HOSTPLANT-HABITATS - 7 X 5 KM ZONE

Fig.102. GRASSLAND, hostplant-habitat for Hesperiids & Satyrines

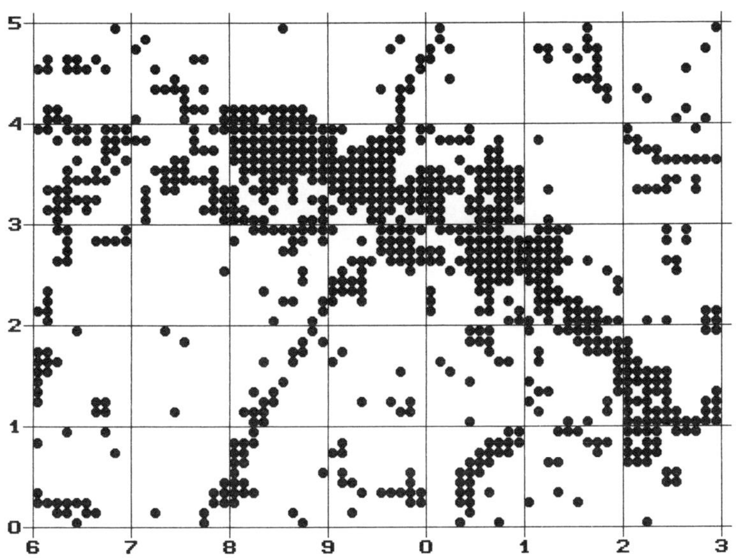

Fig.103. ALDER BUCKTHORN, hostplant-habitat for G.rhamni

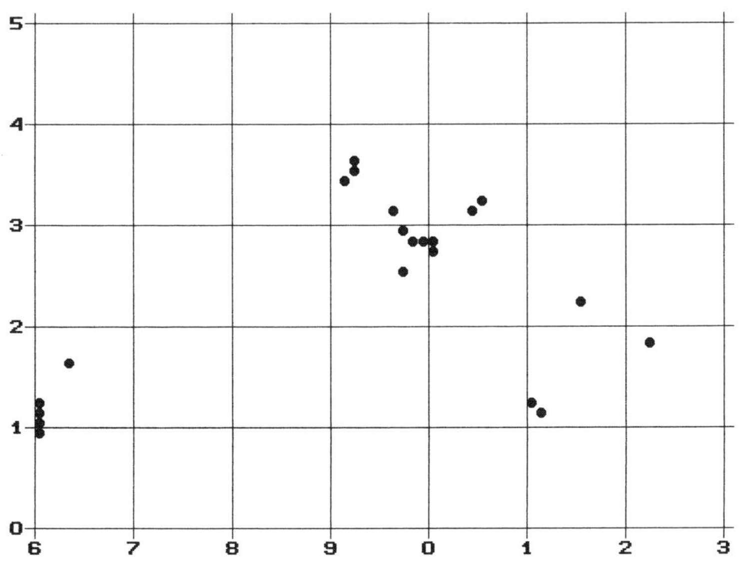

HOSTPLANT-HABITATS - 7 X 5 KM ZONE

Fig.104. CULTIVATED CRUCIFERS, hostplant-habitat for P.brassicae & P.rapae

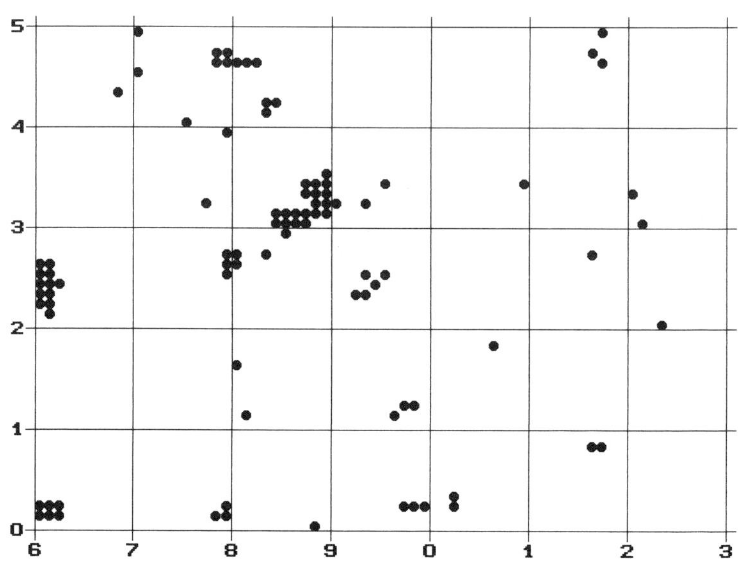

Fig.105. WILD CRUCIFERS, hostplant-habitat for P.brassicae, P.rapae, P.napi

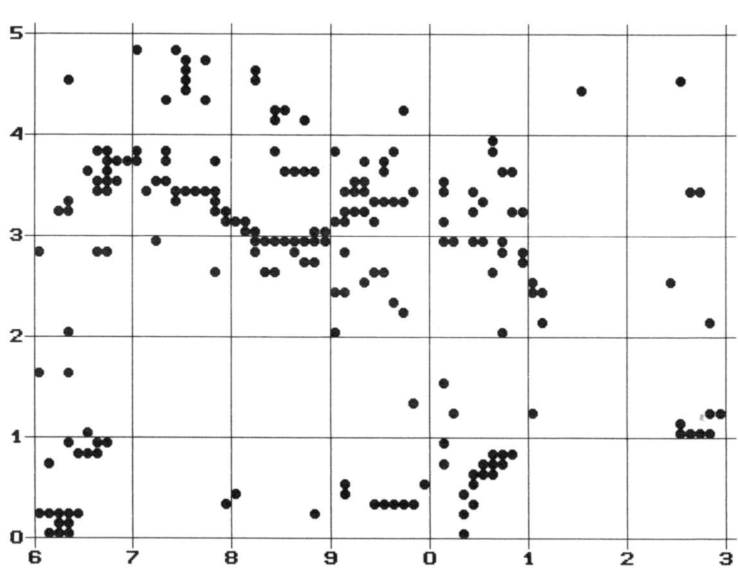

Fig.106. CUCKOO FLOWER, hostplant-habitat for A.cardamines & P.napi

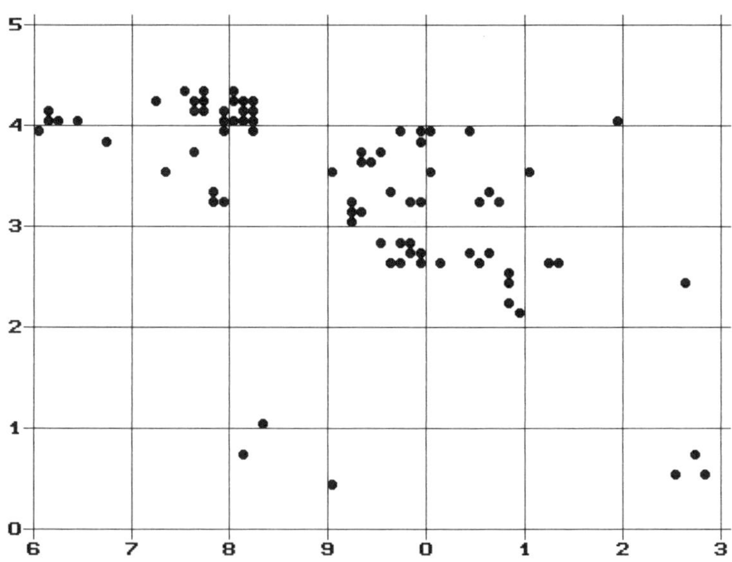

Fig.107. GARLIC MUSTARD, hostplant-habitat for A.cardamines & P.napi

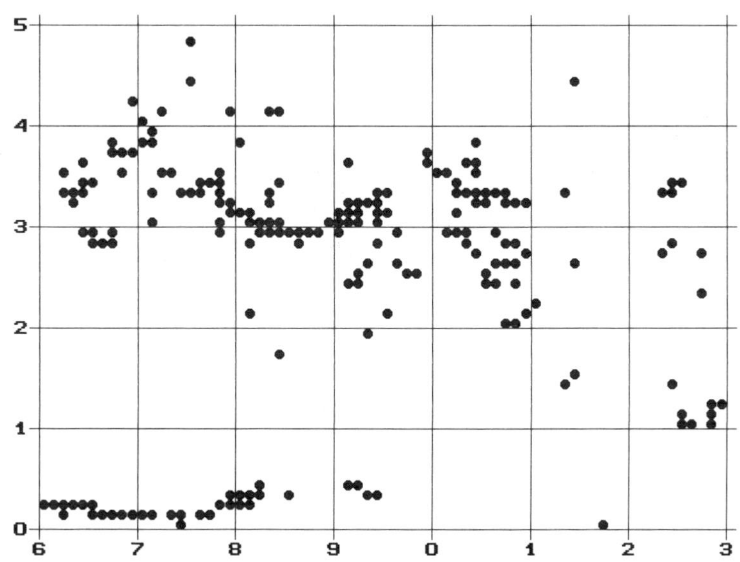

Fig.108. SWEET ROCKET, alternative hostplant for A.cardamines

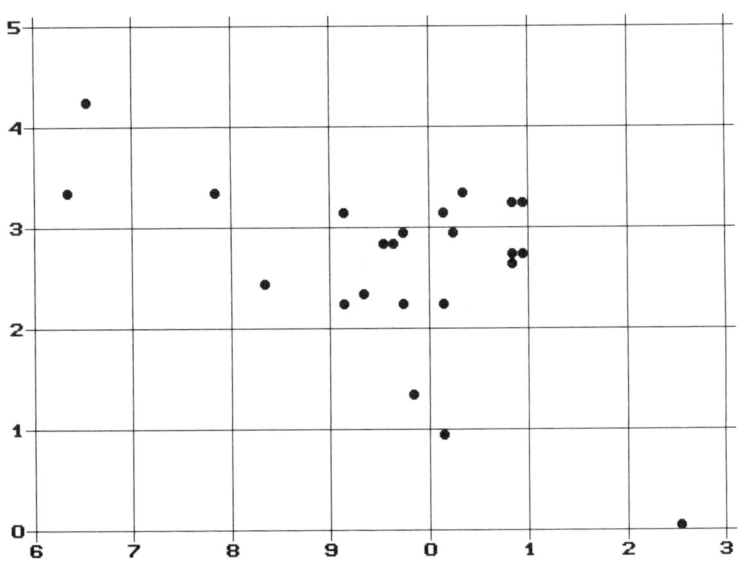

Fig.109. SORREL, hostplant-habitat for L.phlaeas

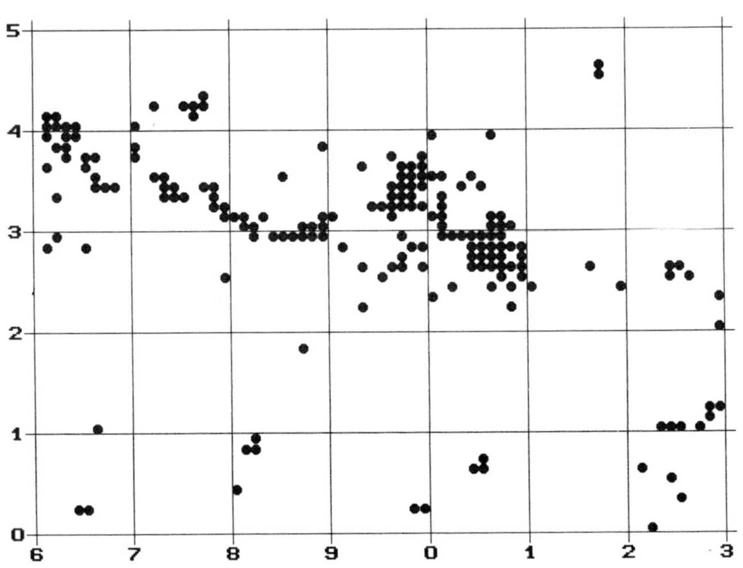

HOSTPLANT-HABITATS - 7 X 5 KM ZONE

Fig.110. CLOVER, hostplant/habitat for C.croceus and P.icarus

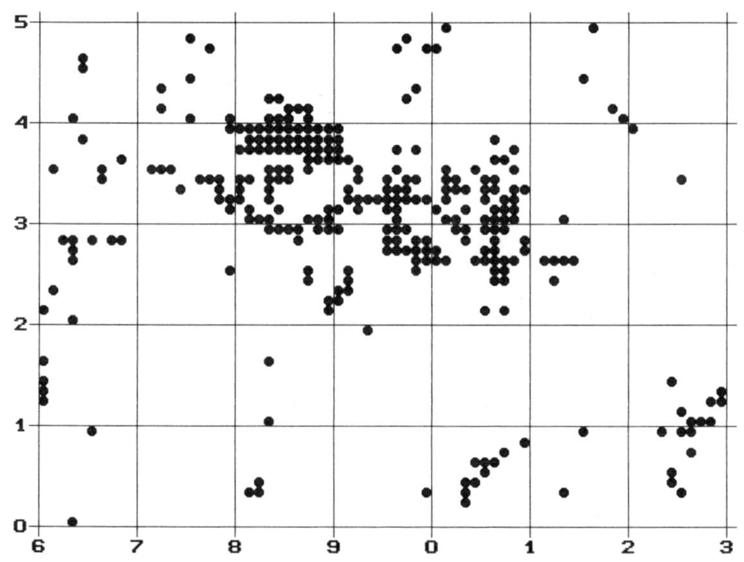

Fig.111. BIRD'S-FOOT TREFOIL, hostplant-habitat for P.icarus

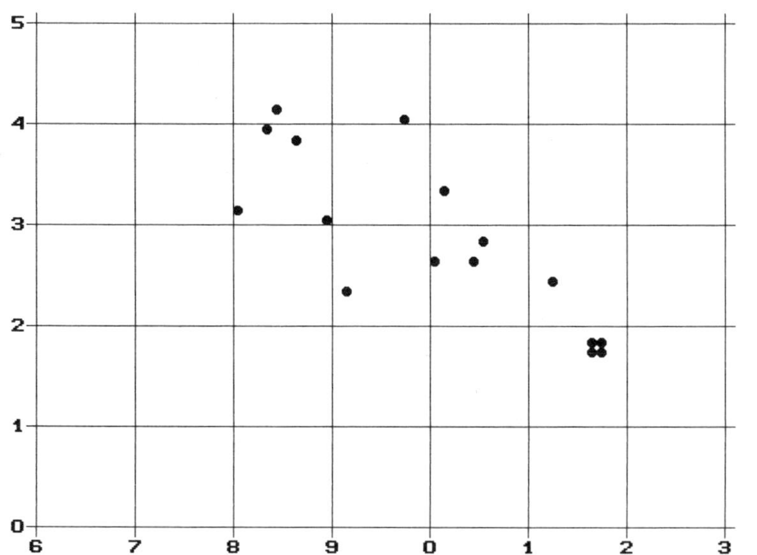

Fig.112. HOLLY, hostplant-habitat for C.argiolus

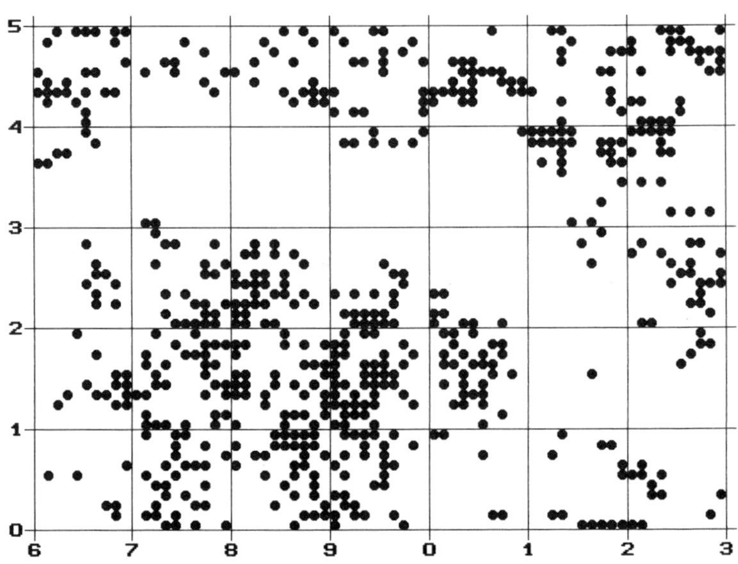

Fig.113. IVY, hostplant-habitat for C.argiolus

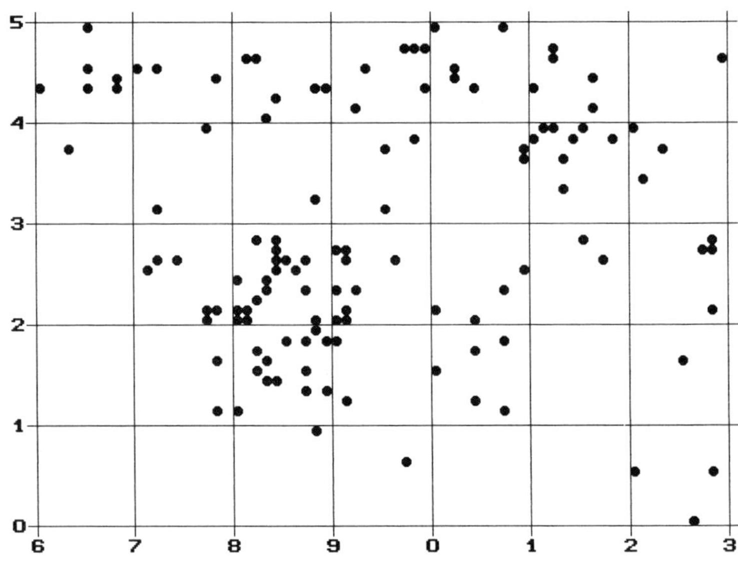

HOSTPLANT-HABITATS - 7 X 5 KM ZONE

ig.114. NETTLE, hostplant-habitat for V.atalanta, A.urticae, I.io, P.c-album

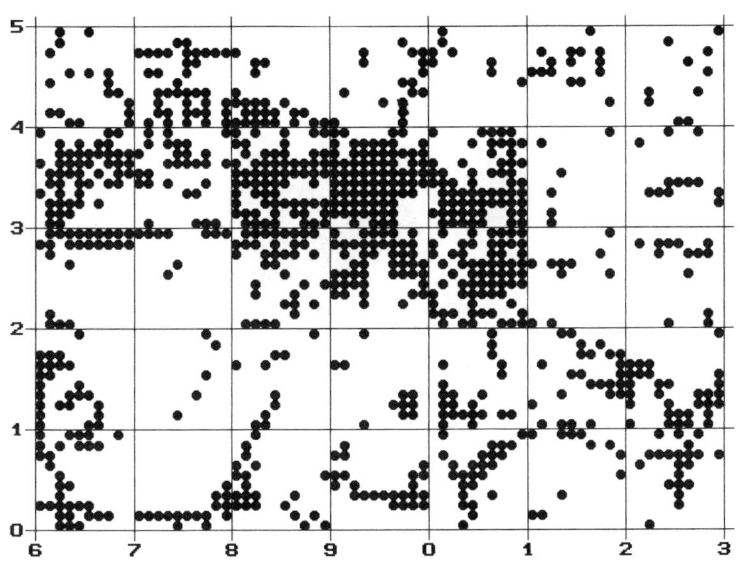

Fig.115. THISTLE, hostplant-habitat for V.cardui

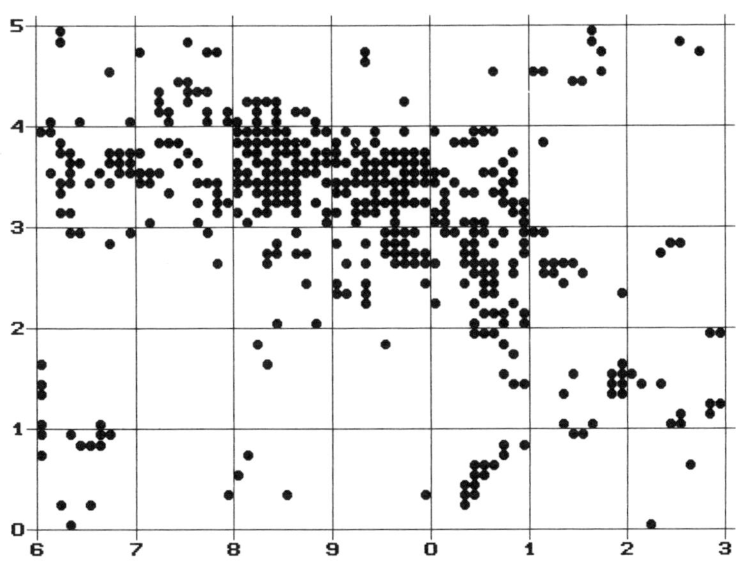

HOSTPLANT-HABITATS - 7 X 5 KM ZONE

Fig.116. WOODLAND WITH GRASS, hostplant-habitat for P.aegeria

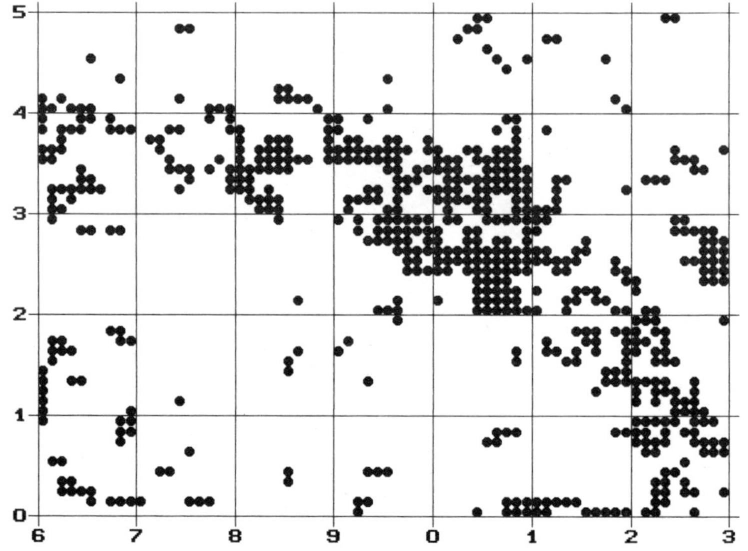

3 X 2 KM ZONE

Fig.117. THE 3 X 2 KM ZONE

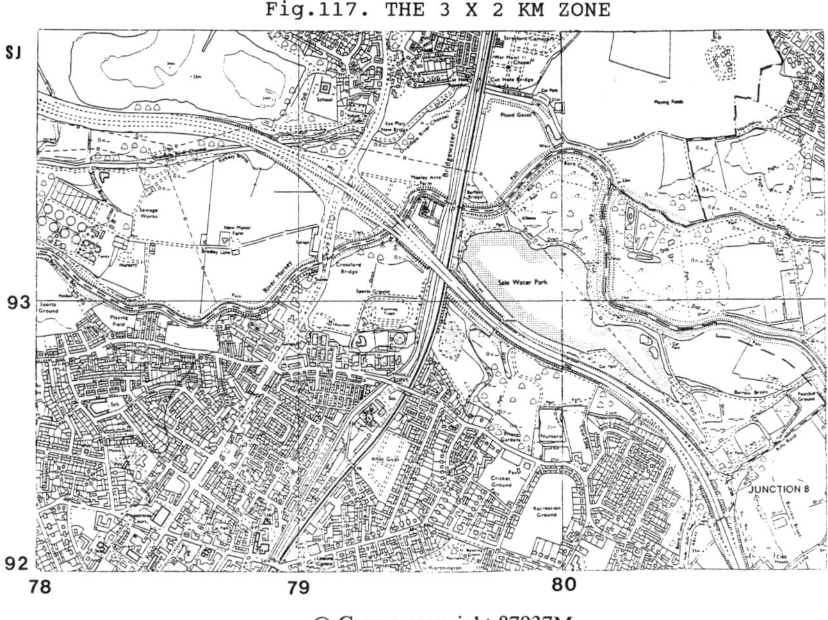

ENVIRONMENTS IN THE 3 X 2 KM ZONE

In figures 118 to 125, the graded symbols denote the approximate percentage of each 100 m square covered by the environment type, thus:

			Nil cover
•	⊗	○	1-5 % cover
●	⊗	○	6-10% cover
●	⊗	○	11-20% cover
●	⊗	○	21-50% cover
●	⊗	○	51-75% cover
●	⊗	○	76-100% cover

The above symbols also apply to the Hostplant-habitat maps, figures 148 to 162.

ENVIRONMENTS - 3 X 2 KM ZONE
For explanation of symbols see page 101

Fig.118. GRASSLAND ASSESSED AS POTENTIALLY SUITABLE HABITAT

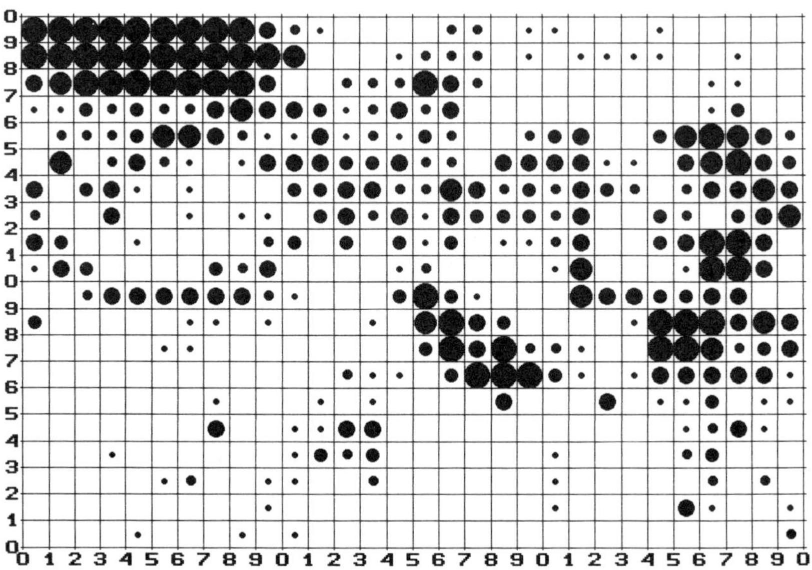

Fig.119. WOODLAND AND SCRUB

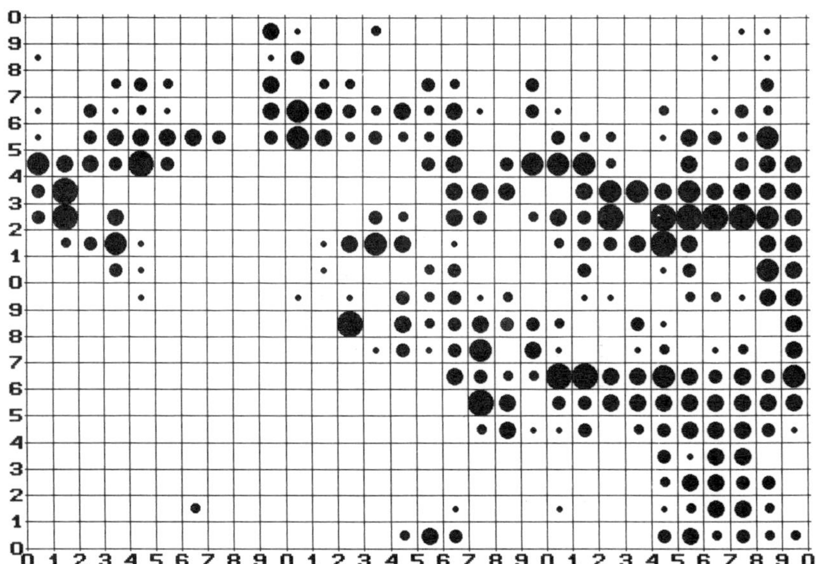

ENVIRONMENTS - 3 X 2 KM ZONE

For explanation of symbols see page 101

Fig.120. OPEN SPACES UNSUITABLE AS HABITAT

Black, heavily grazed grass. Circles/crosses, mown grass. Open circles, marsh & long grass

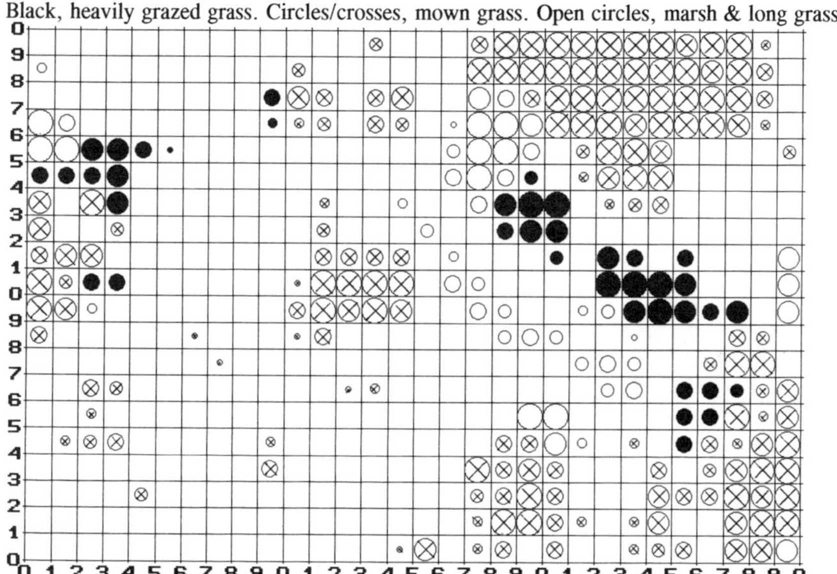

Fig.121. FARMLAND AND ALLOTMENTS

Black, farmland. Open circles, allotments.

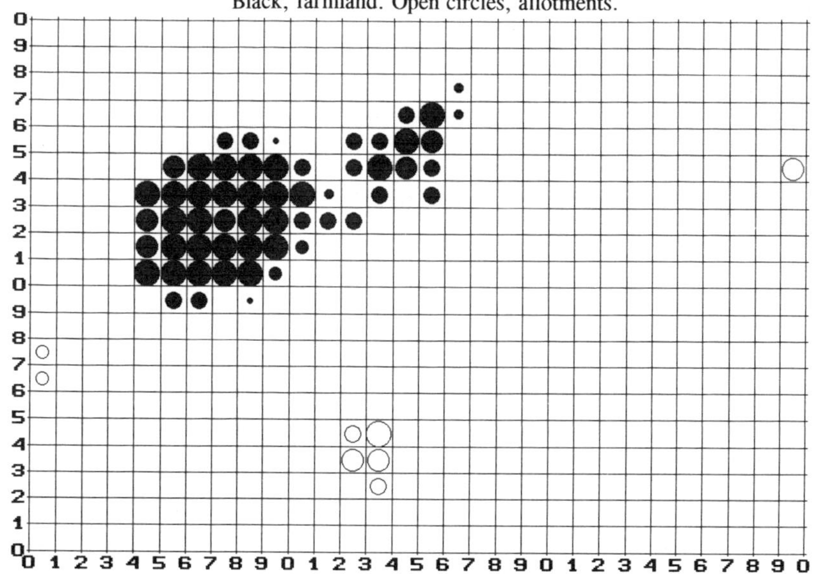

ENVIRONMENTS - 3 X 2 KM ZONE

For explanation of symbols see page 101

Fig.122. URBAN COVER - HOUSES AND GARDENS

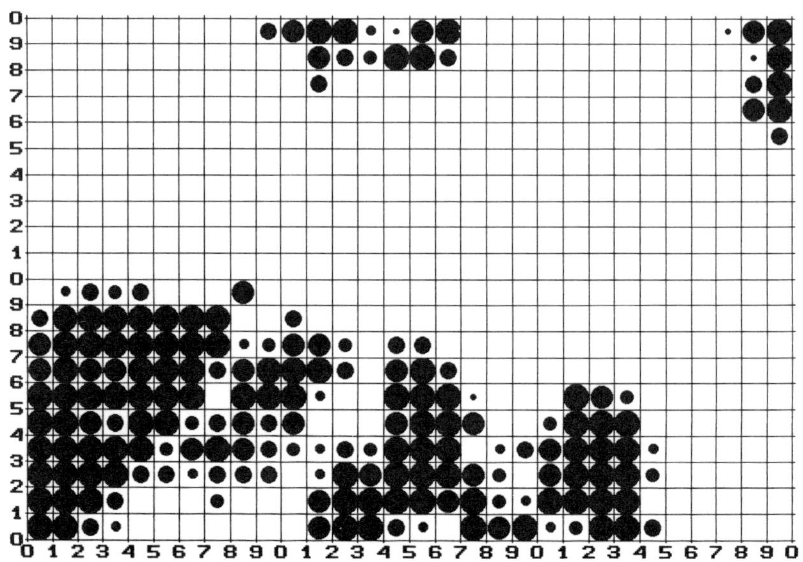

Fig.123. URBAN COVER - OFFICES, SHOPS AND INDUSTRIAL AREAS

Open circles, offices, shops & industry. Black symbols, sewage works.

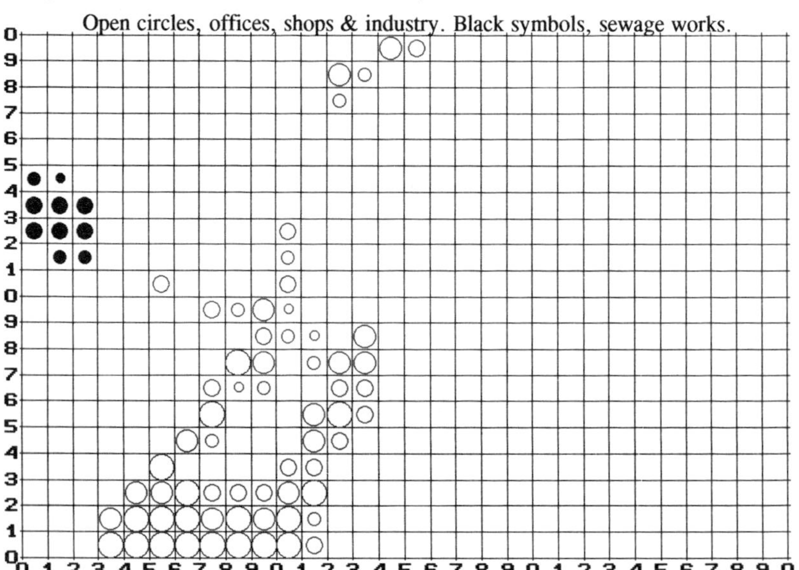

ENVIRONMENTS - 3 X 2 KM ZONE

For explanation of symbols see page 101

Fig.124. URBAN COVER - MAIN ROADS

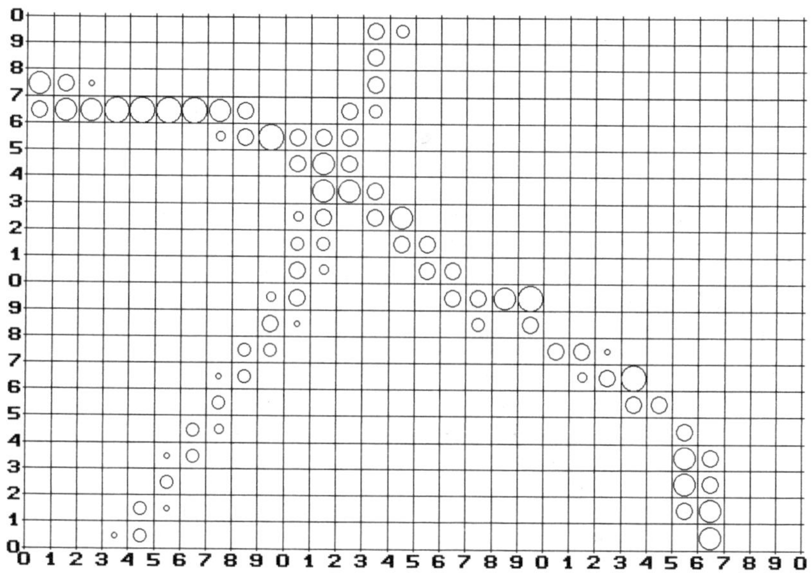

Fig.125. WATER - LAKES, RIVER AND CANAL

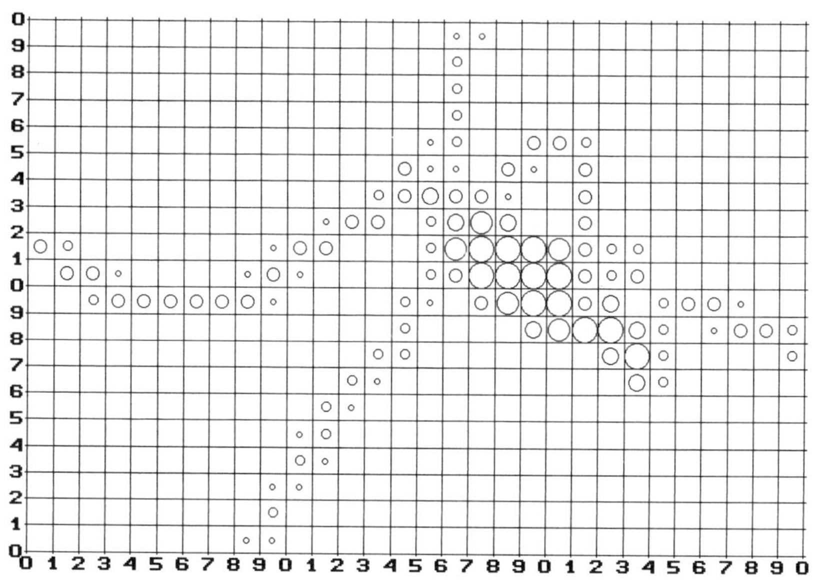

RECORDING COVERAGE - 3 X 2 KM ZONE

Fig.126. RECORDING COVERAGE 1994-7, DURING BUTTERFLY SEASON

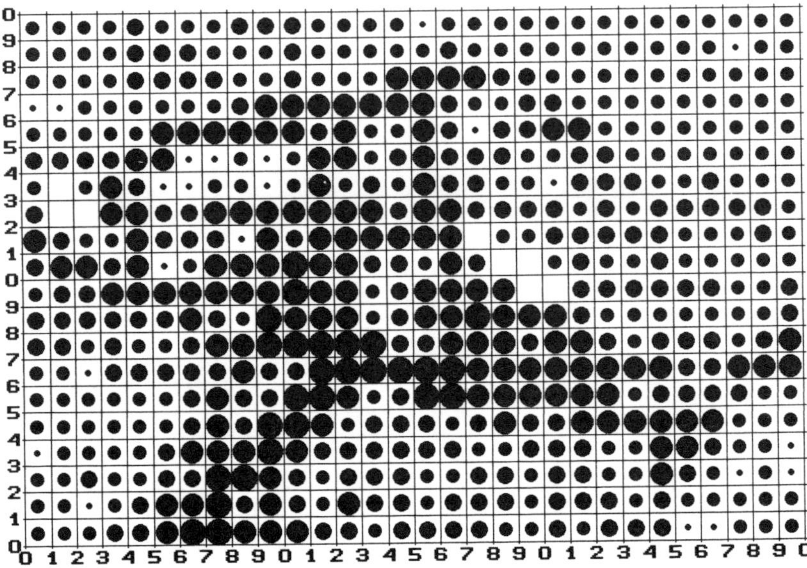

The symbols in the above figure represent the number of visits to each square during the four "butterfly seasons", thus:

- not visited - see page 6
- 1 visit
- 2-9 visits
- 10-29 visits
- 30-99 visits
- 100+ visits

Fig.127. RECORDS FOR ALL SPECIES IN 3 X 2 KM ZONE

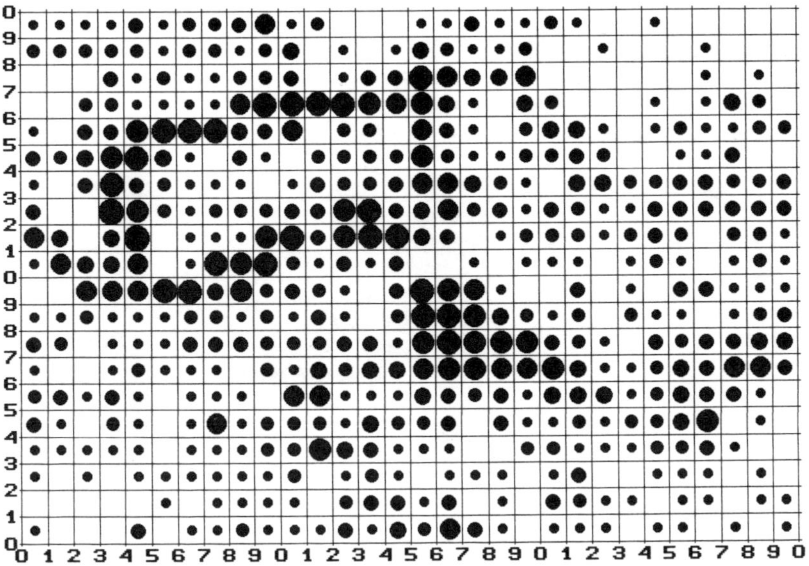

NUMBERS OF SPECIES RECORDED IN EACH 100 M SQUARE

1	2	2	2	6	2	4	3	6	9	2	3	0	0	0	1	2	5	1	2	4	1	0	0	2	0	0	0	0	0
3	3	3	3	4	2	3	4	2	3	8	0	2	0	1	7	4	1	2	4	0	0	1	0	0	0	1	0	0	0
0	0	0	6	2	3	1	2	3	4	6	0	1	6	6	15	9	8	8	9	0	0	0	0	0	0	1	0	2	0
0	0	4	3	2	1	1	2	9	16	13	13	16	11	9	11	7	2	0	7	3	0	0	0	1	0	2	7	4	0
1	0	6	6	11	13	14	13	8	5	9	0	4	4	0	9	5	1	0	4	7	8	2	0	2	4	2	2	4	3
5	3	7	11	15	8	2	0	5	1	0	3	4	4	4	11	4	2	2	3	3	5	3	0	0	1	1	6	0	0
2	0	5	13	5	3	1	1	1	0	1	6	4	4	6	9	9	8	4	1	0	7	8	3	3	5	6	3	3	6
6	0	0	18	11	4	2	4	4	3	5	5	12	15	5	8	10	4	6	2	6	4	1	2	6	6	5	6	6	6
9	7	0	8	17	0	1	1	1	12	15	5	11	13	13	8	5	0	1	3	4	3	1	4	5	4	0	4	4	2
2	10	8	8	10	0	1	13	11	15	3	1	6	1	5	0	0	2	0	2	1	2	0	2	3	2	0	1	4	1
0	0	9	9	10	14	15	7	11	3	6	3	1	0	6	15	11	11	1	0	0	5	0	1	0	6	5	2	2	1
2	1	4	2	1	1	4	3	2	3	2	5	2	0	3	13	13	12	8	3	2	3	0	3	1	1	0	2	3	5
5	3	0	1	1	1	5	5	5	4	4	4	5	5	2	11	13	12	12	11	6	4	1	0	5	3	6	5	7	8
2	0	0	1	3	0	2	2	0	3	1	8	3	7	8	9	12	12	9	9	12	8	2	2	3	7	6	9	10	5
4	5	1	5	1	0	2	1	2	0	10	10	1	1	2	8	6	4	5	2	7	8	7	2	3	7	6	6	1	0
6	1	0	4	1	0	1	10	2	3	4	4	1	7	4	4	5	0	6	2	2	3	2	3	4	8	11	0	2	0
2	2	1	2	1	0	1	2	1	3	4	11	7	6	2	1	1	0	0	3	3	1	1	1	3	4	6	1	0	0
1	0	2	0	1	1	2	2	2	2	3	0	1	4	1	0	1	2	1	0	2	5	0	0	2	1	2	0	2	0
0	0	0	0	0	1	0	1	1	1	1	0	4	6	6	2	5	0	1	0	5	3	1	1	0	2	1	0	1	1
1	0	0	0	5	0	1	1	3	1	1	2	5	1	8	4	9	5	1	0	2	1	1	0	2	1	0	1	0	2

BUTTERFLY DISTRIBUTIONS, 1992-1997 - 3 X 2 KM ZONE

Small symbols, single sightings. Full-size symbols, multiple sightings.

Fig.128. Thymelicus sylvestris

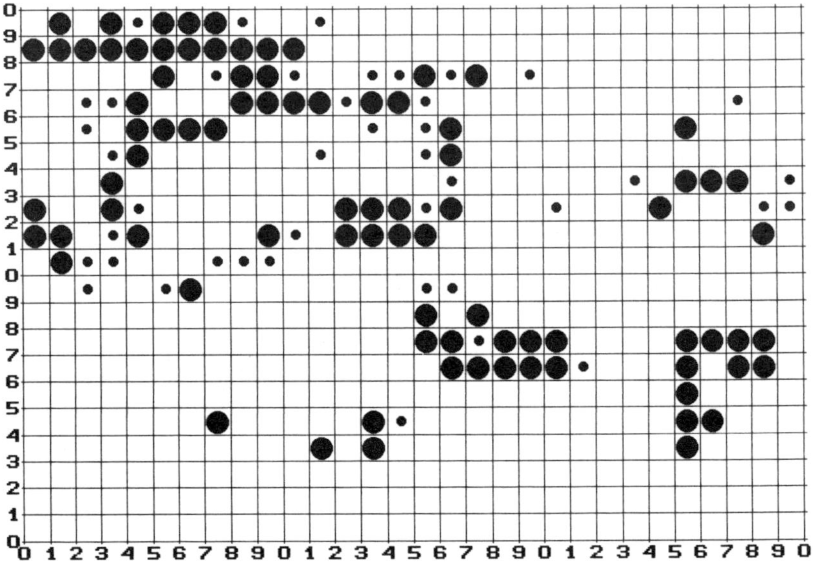

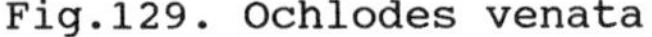

Fig.129. Ochlodes venata

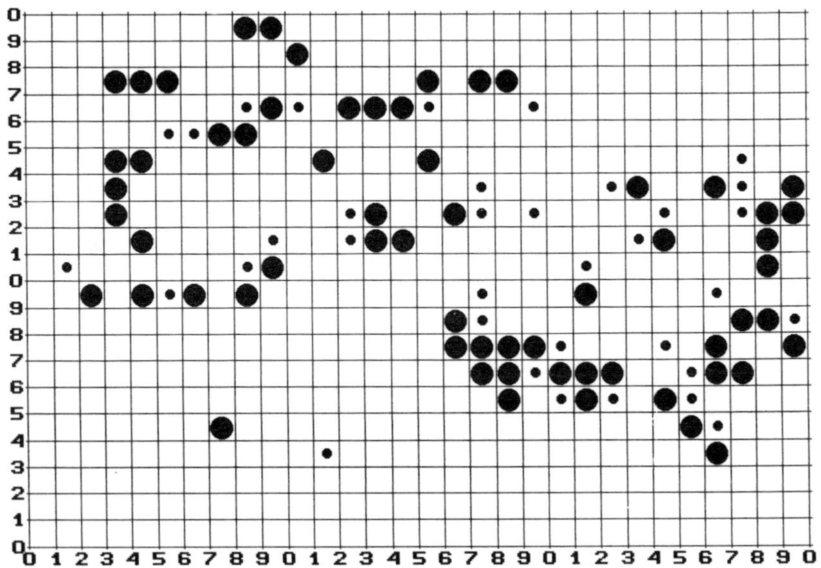

BUTTERFLY DISTRIBUTIONS, 1992-1997 - 3 X 2 KM ZONE
Small symbols, single sightings. Full-size symbols, multiple sightings.

Fig.130. Colias croceus

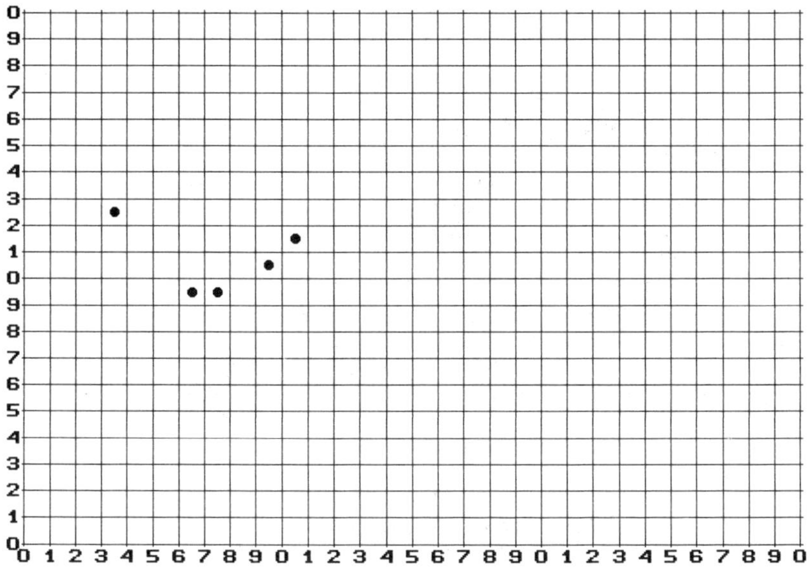

0 9 8 7 6 5 4 3 2 1 0 9 8 7 6 5 4 3 2 1 0
0 1 2 3 4 5 6 7 8 9 0 1 2 3 4 5 6 7 8 9 0 1 2 3 4 5 6 7 8 9 0

Fig.131. Gonepteryx rhamni

0 9 8 7 6 5 4 3 2 1 0 9 8 7 6 5 4 3 2 1 0
0 1 2 3 4 5 6 7 8 9 0 1 2 3 4 5 6 7 8 9 0 1 2 3 4 5 6 7 8 9 0

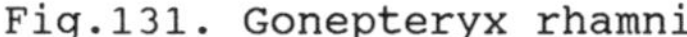

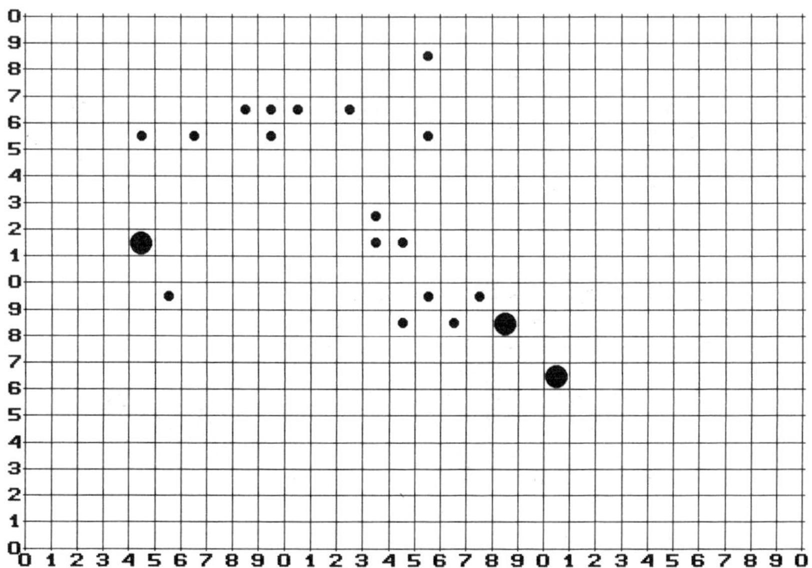

BUTTERFLY DISTRIBUTIONS, 1992-1997 - 3 X 2 KM ZONE

Small symbols, single sightings. Full-size symbols, multiple sightings.

Fig.132. Pieris brassicae

Fig.133. Pieris rapae

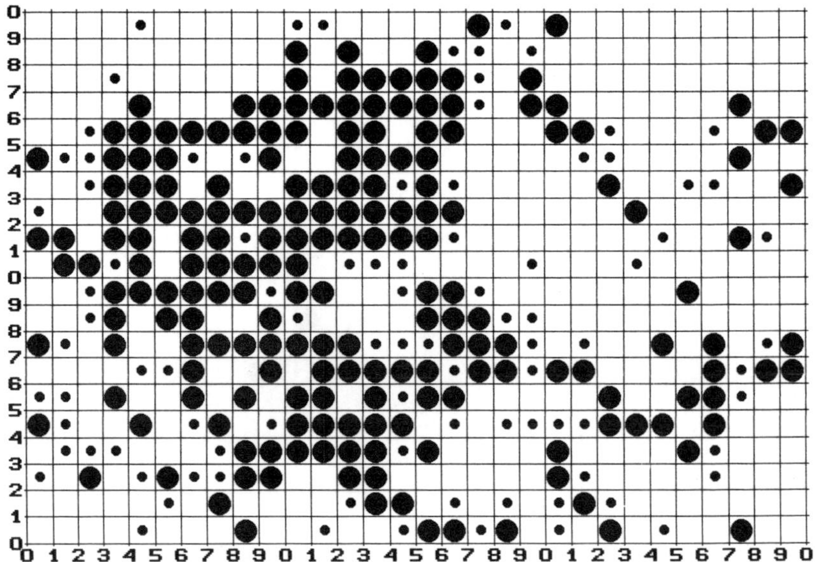

BUTTERFLY DISTRIBUTIONS, 1992-1997 - 3 X 2 KM ZONE

Small symbols, single sightings. Full-size symbols, multiple sightings.

Fig.134. Pieris napi

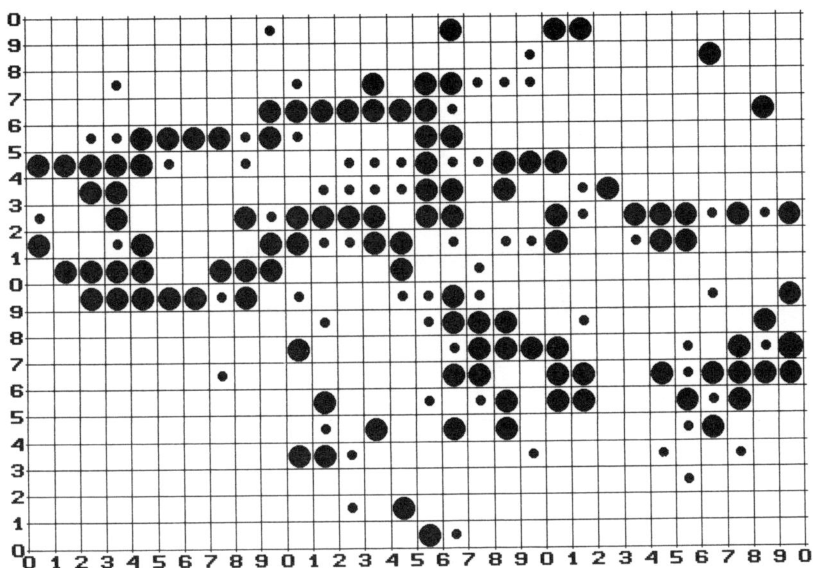

Fig.135. Anthocharis cardamines

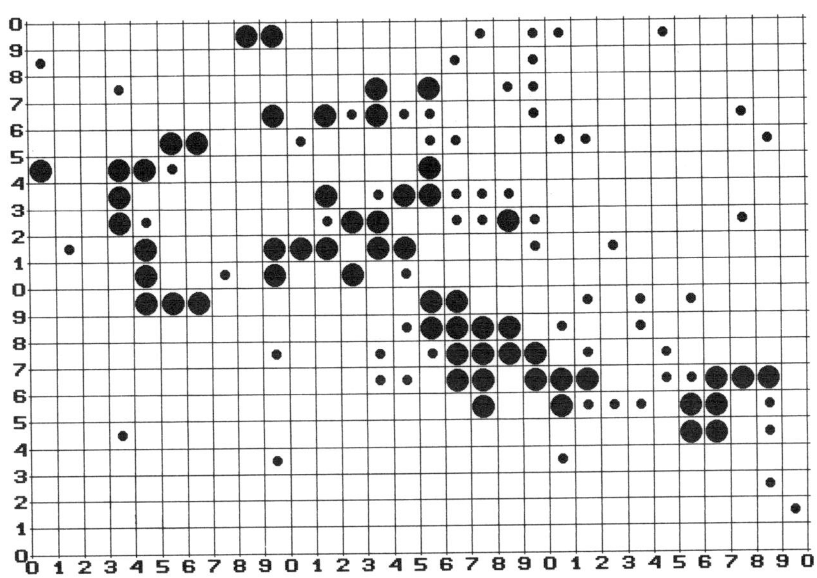

BUTTERFLY DISTRIBUTIONS, 1992-1997 - 3 X 2 KM ZONE

Small symbols, single sightings, Full-size symbols, multiple sightings.
Open circle, single larval record in 1992 (see page 27).

Fig.136. Lycaena phlaeas

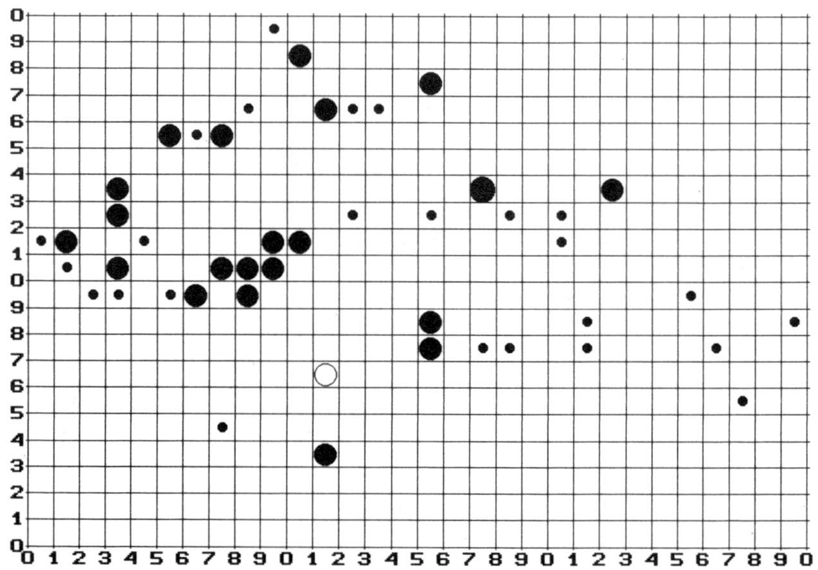

Fig.137. Polyommatus icarus

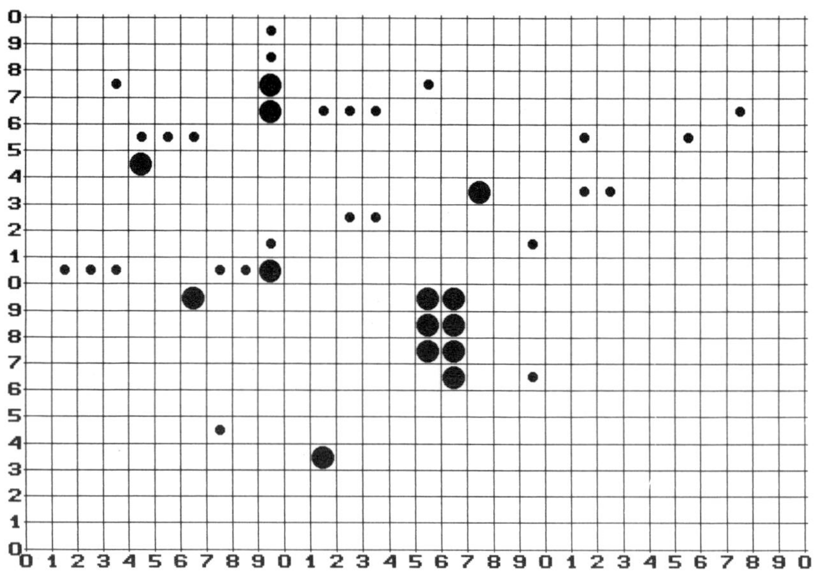

BUTTERFLY DISTRIBUTIONS, 1992-1997 - 3 X 2 KM ZONE

Small symbols, single sightings. Full-size symbols, multiple sightings.

Fig.138. Celastrina argiolus

0 9 8 7 6 5 4 3 2 1 0 9 8 7 6 5 4 3 2 1 0

0 1 2 3 4 5 6 7 8 9 0 1 2 3 4 5 6 7 8 9 0 1 2 3 4 5 6 7 8 9 0

Fig.139. Vanessa atalanta

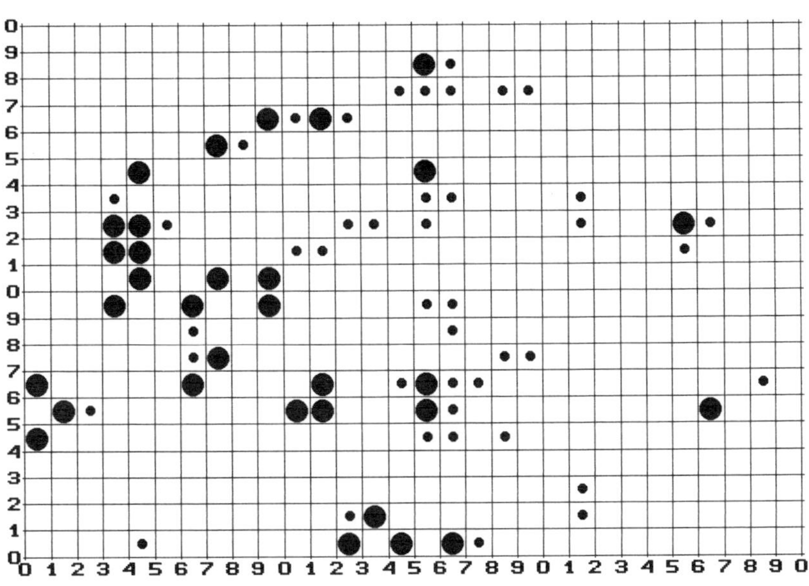

BUTTERFLY DISTRIBUTIONS, 1992-1997 - 3 X 2 KM ZONE

Small symbols, single sightings. Full-size symbols, multiple sightings.

Fig.140. Vanessa cardui

Fig.141. Aglais urticae

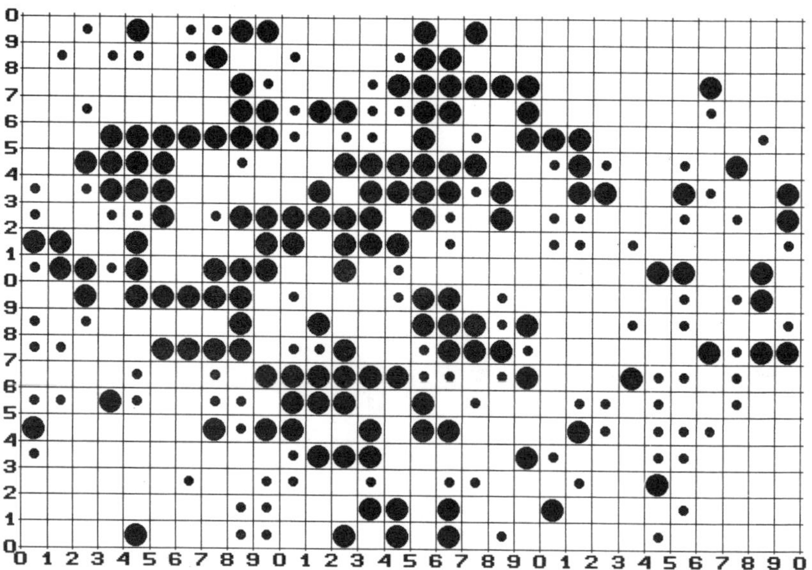

BUTTERFLY DISTRIBUTIONS, 1992-1997 - 3 X 2 KM ZONE
Small symbols, single sightings. Full-size symbols, multiple sightings.

Fig.142. Inachis io

Fig.143. Polygonia c-album

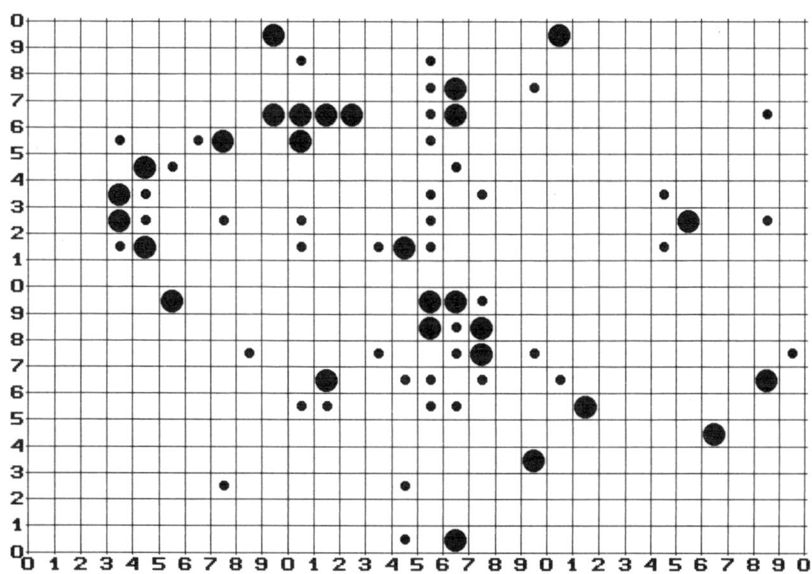

BUTTERFLY DISTRIBUTIONS, 1992-1997 - 3 X 2 KM ZONE

Small symbols, single sightings. Full-size symbols, multiple sightings.

Fig.144. Pararge aegeria

Fig.145. Lasiommata megera

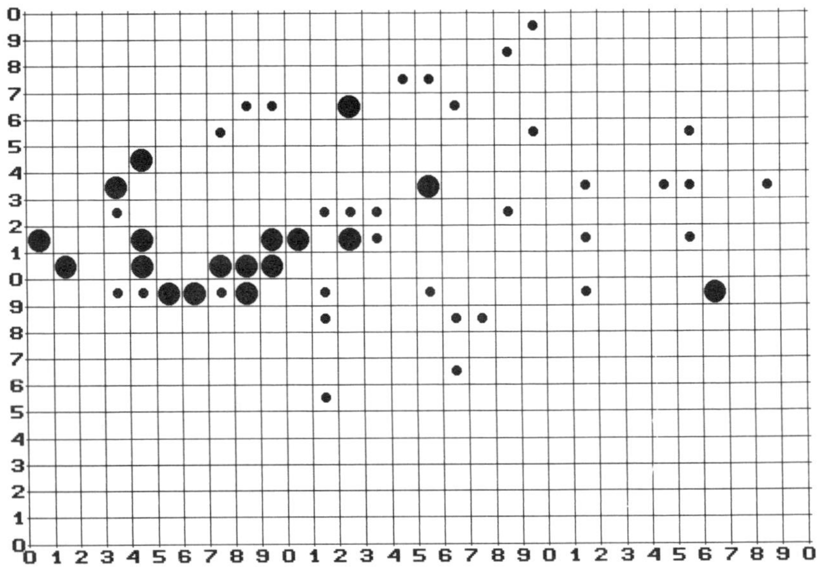

BUTTERFLY DISTRIBUTIONS, 1992-1997 - 3 X 2 KM ZONE
Small symbols, single sightings. Full-size symbols, multiple sightings.

Fig.146. Pyronia tithonus

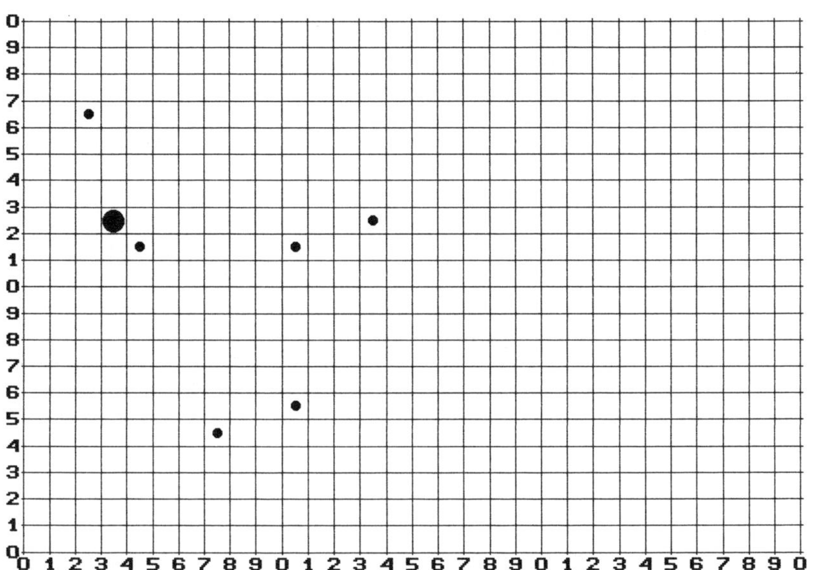

Fig.147. Maniola jurtina

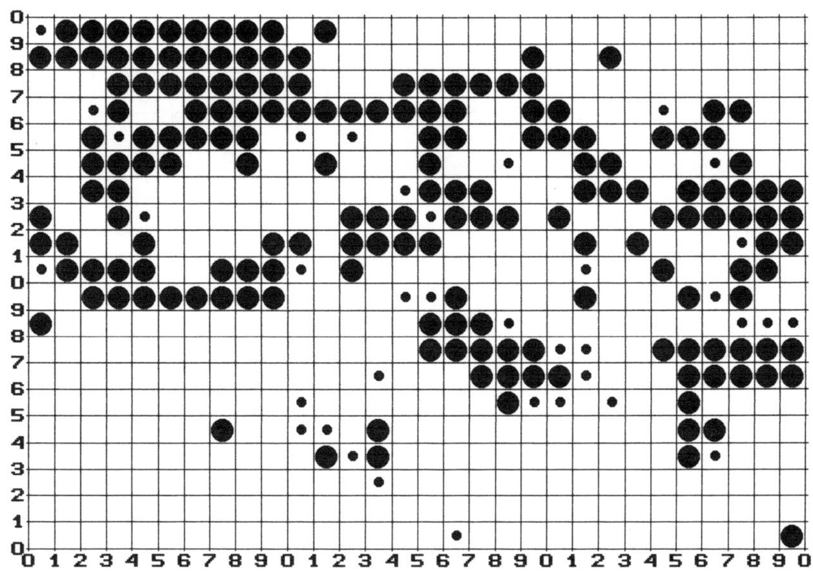

HOSTPLANT-HABITATS, 3 X 2 KM ZONE

Symbol sizes, same as for Environments - see page 101

Fig.148. GRASSLAND, hostplant-habitat for Hesperiids and Satyrines

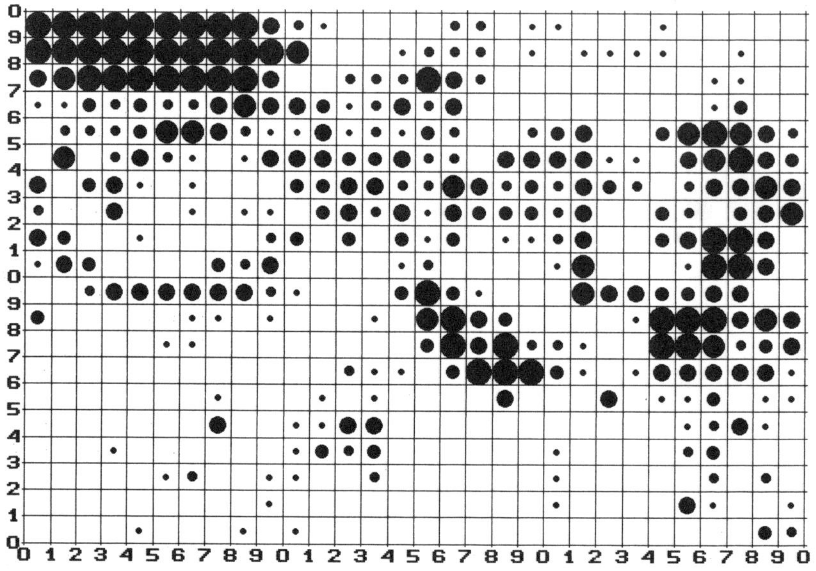

Fig.149. ALDER BUCKTHORN, hostplant-habitat for G.rhamni

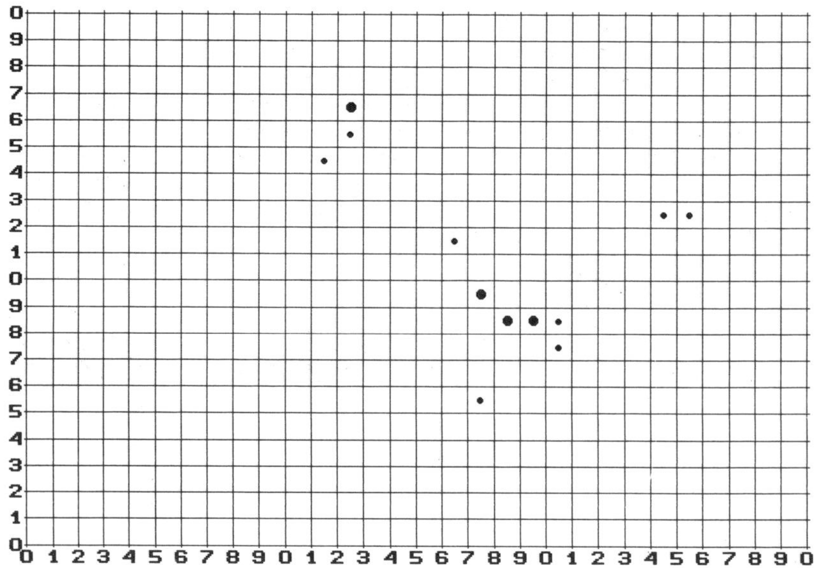

HOSTPLANT-HABITATS, 3 X 2 KM ZONE

Symbol sizes, same as for Environments - see page 101

Fig.150. CULTIVATED CRUCIFERS, hostplant-habitat for P.brassicae & P.rapae

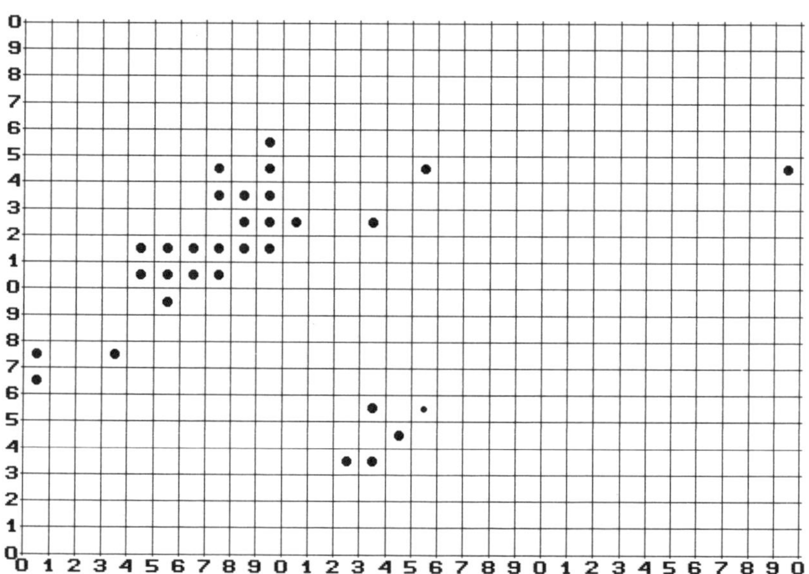

Fig.151. WILD CRUCIFERS, hostplant-habitat for P.brassicae, P.rapae, P.napi

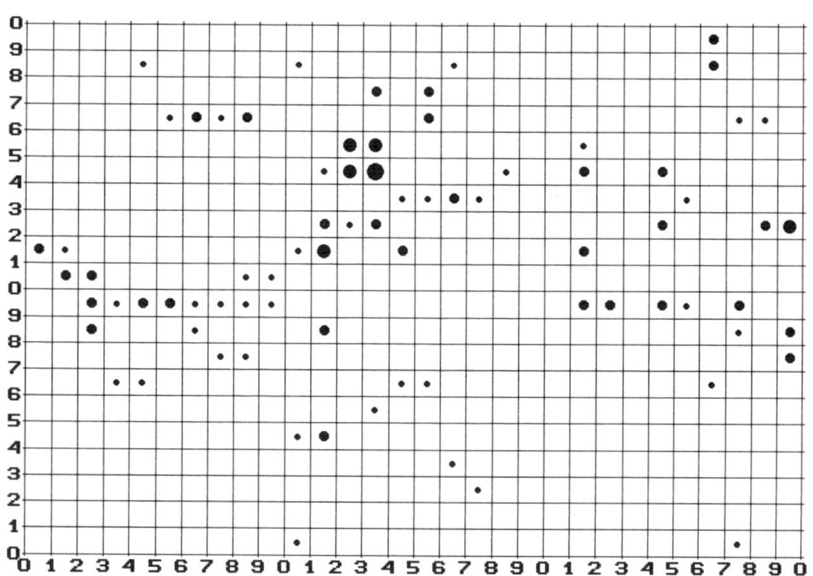

HOSTPLANT-HABITATS, 3 X 2 KM ZONE

Symbol sizes, same as for Environments - see page 101

Fig.152. CUCKOO FLOWER, hostplant-habitat for P.napi & A.cardamines

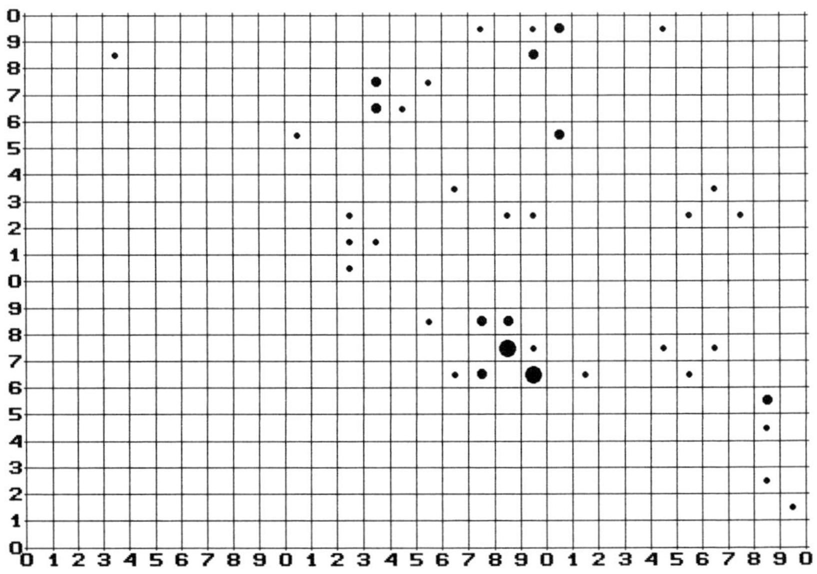

Fig.153. GARLIC MUSTARD, hostplant-habitat for P.napi & A.cardamines

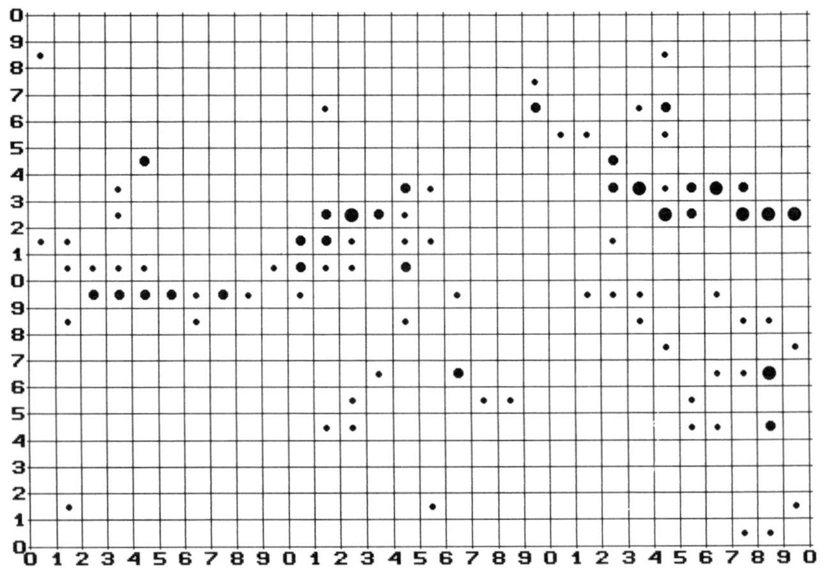

HOSTPLANT-HABITATS, 3 X 2 KM ZONE

Symbol sizes, same as for Environments - see page 101

Fig.154. HONESTY & SWEET ROCKET, alternative hostplants for A.cardamines

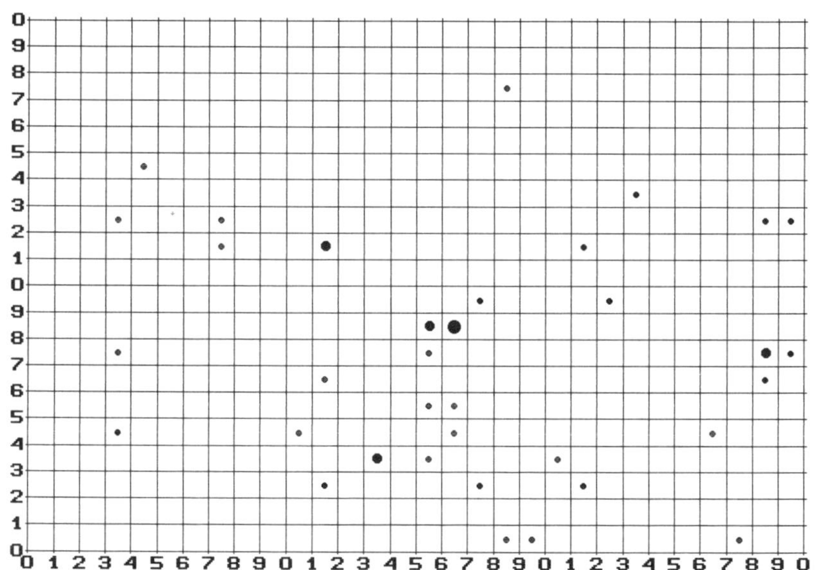

Fig.155. SORREL, hostplant-habitat for L.phlaeas

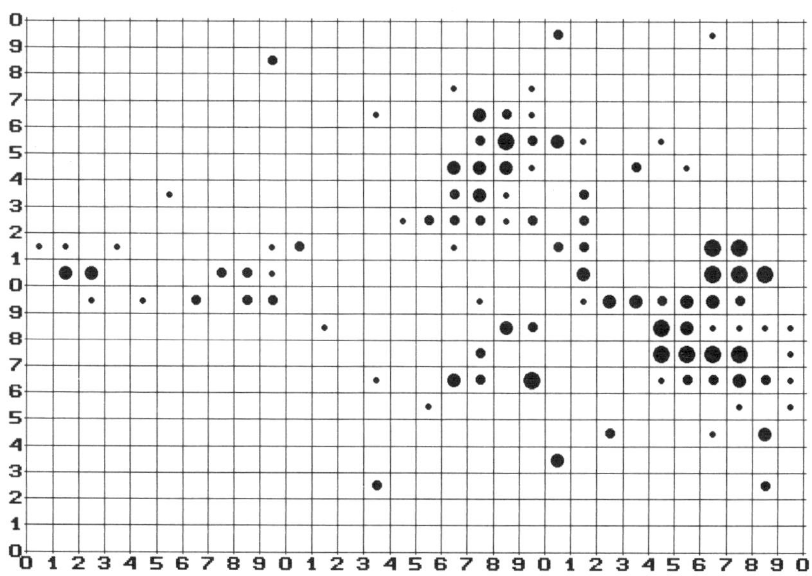

HOSTPLANT-HABITATS, 3 X 2 KM ZONE

Symbol sizes, same as for Environments - see page 101

Fig.156 CLOVER, hostplant-habitat for C.croceus and P.icarus

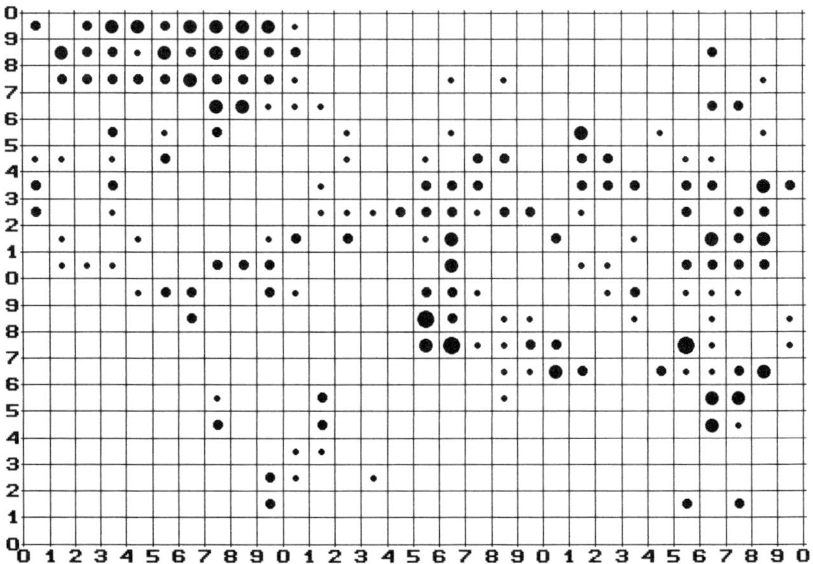

Fig.157.BIRD'S-FOOT TREFOIL, hostplant-habitat for P.icarus

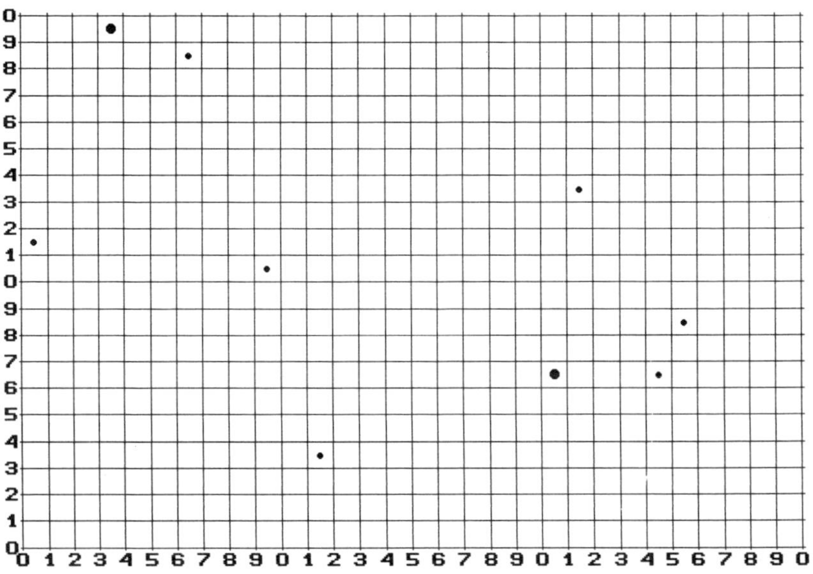

HOSTPLANT-HABITATS, 3 X 2 KM ZONE

Symbol sizes, same as for Environments - see page 101

Fig.158. HOLLY, hostplant-habitat for C.argiolus

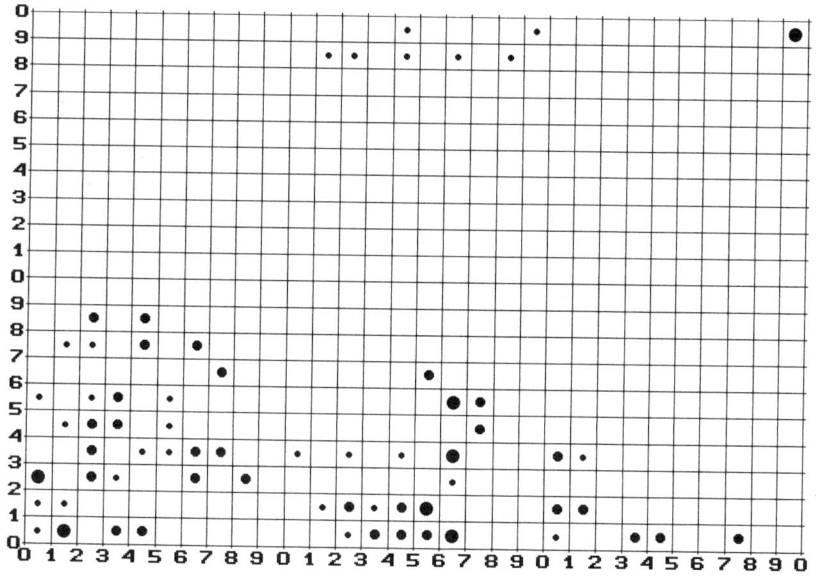

Fig.159. IVY, hostplant-habitat for C.argiolus

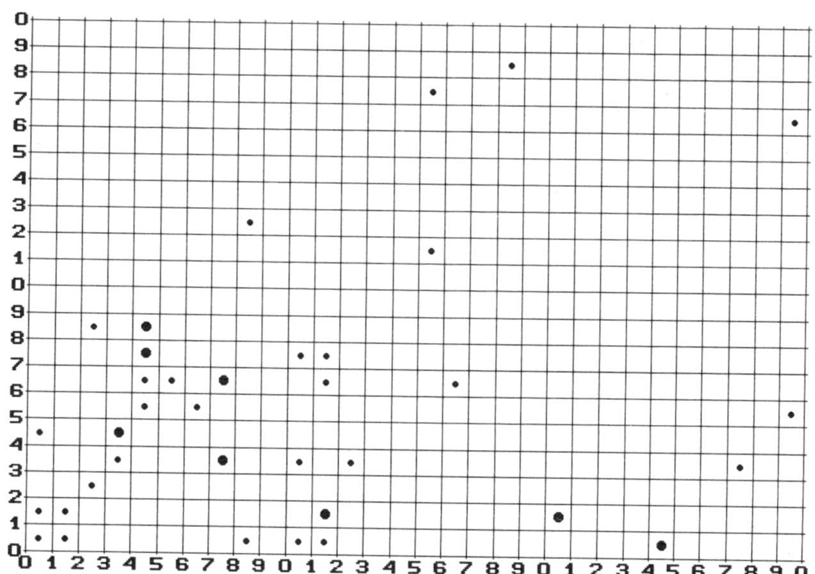

Symbol sizes, same as for Environments - see page 101

Fig.160. NETTLE, hostplant-habitat for V.atalanta, A.urticae, I.io, P.c-alb

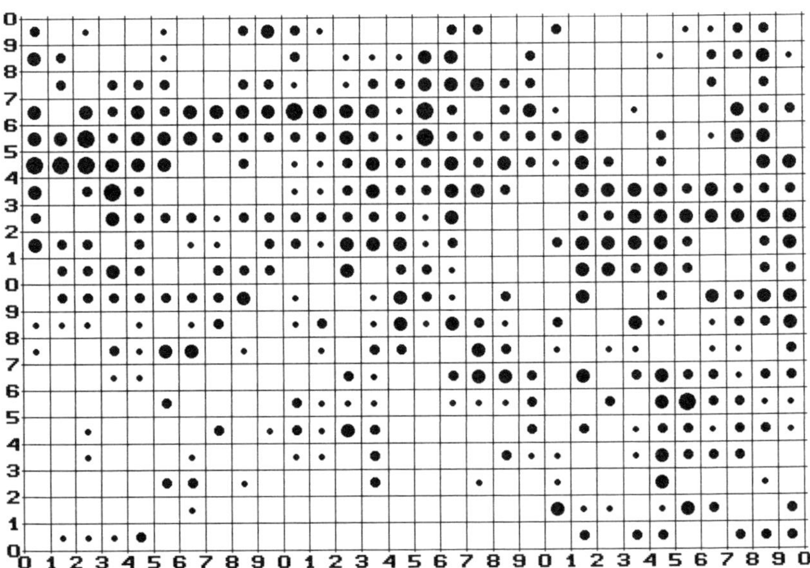

Fig.161. THISTLE, hostplant-habitat for V.cardui

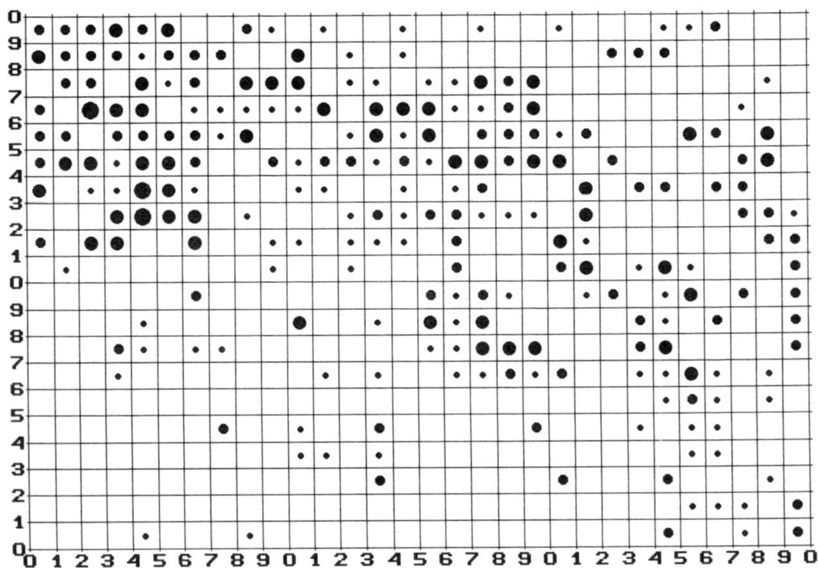

HOSTPLANT-HABITATS, 3 X 2 KM ZONE

Symbol sizes, same as for Environments - see page 101

Fig.162. WOODLAND WITH GRASS, hostplant-habitat for P.aegeria

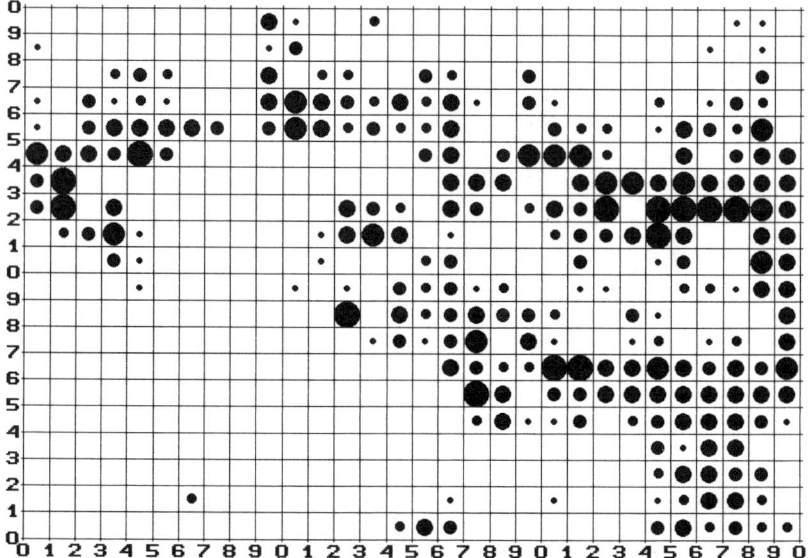

MAIN NECTAR SOURCES, 1996-1997 - 3 X 2 KM ZONE

Fig.163. THISTLE

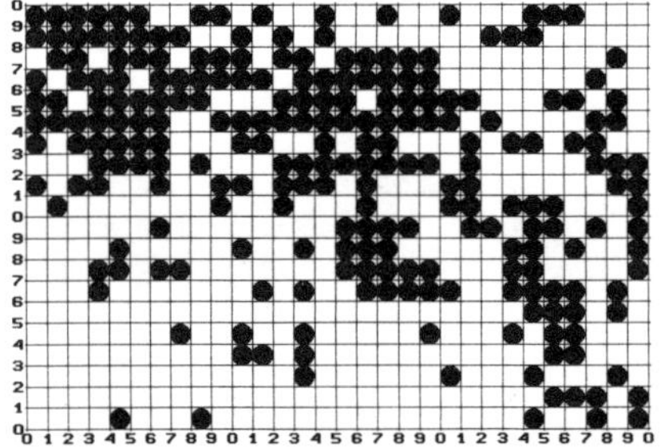

Fig.164. BUDDLEIA

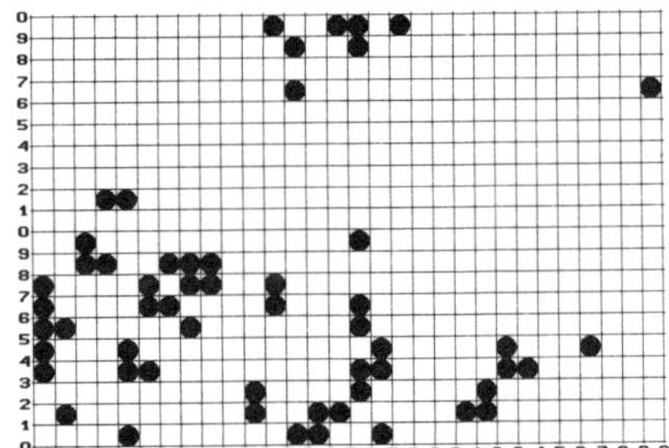

Fig.165. RAGWORT

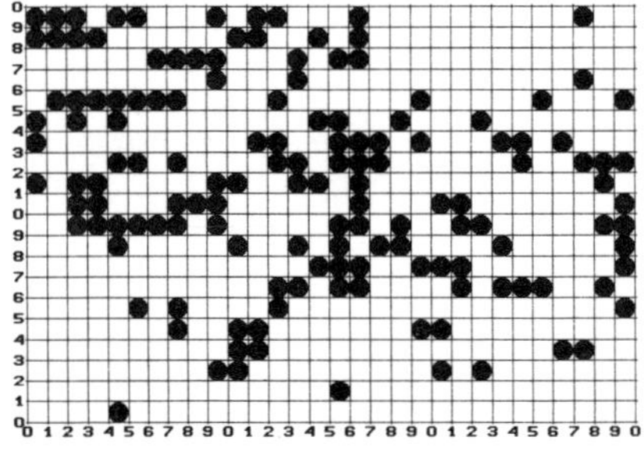

Fig.166. BRAMBLE

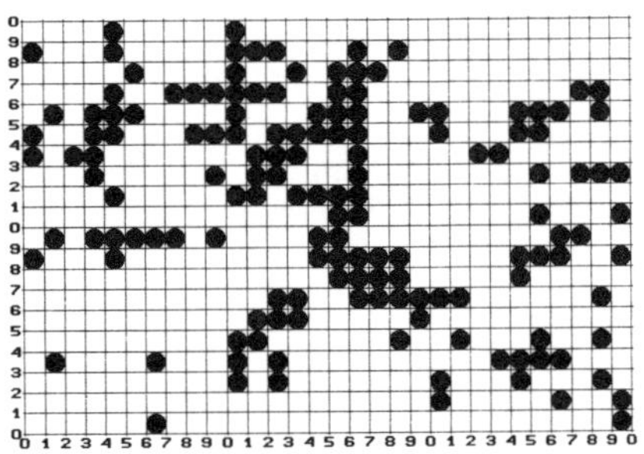

Fig.167. MICHAELMAS DAISY

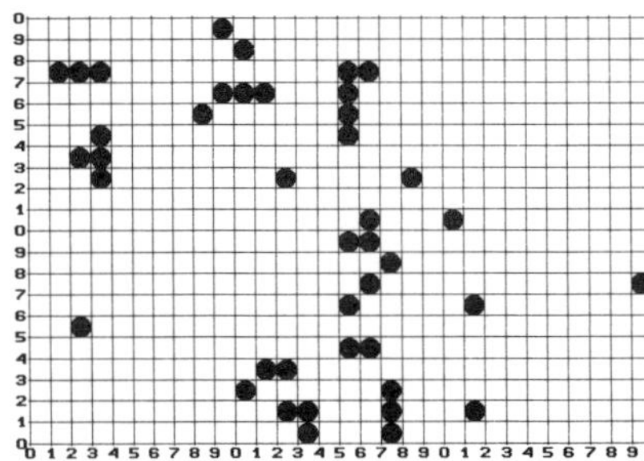

Fig.168. DANDELION

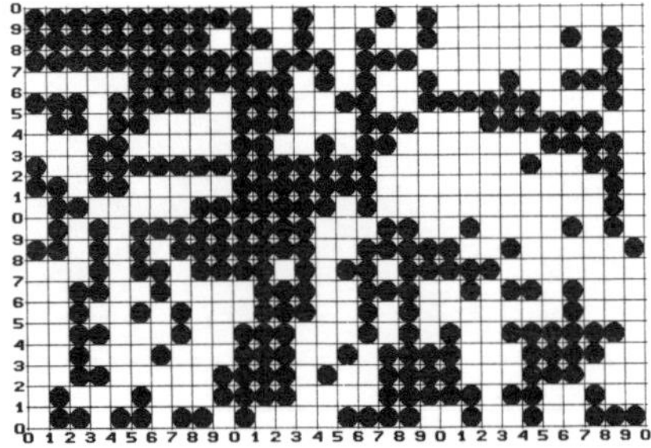

BURNET MOTHS - WHOLE AREA (page 45)

Fig.169. Zygaena lonicerae

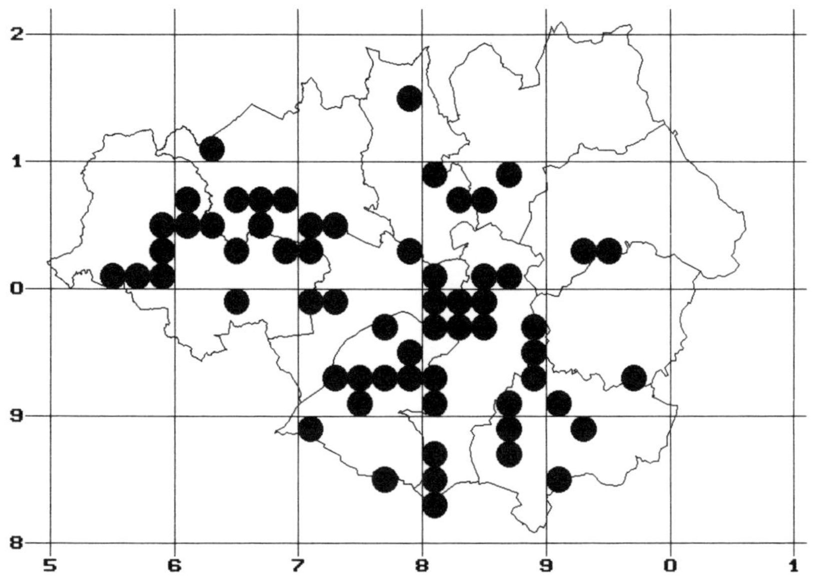

Fig.170. Zygaena filipendulae

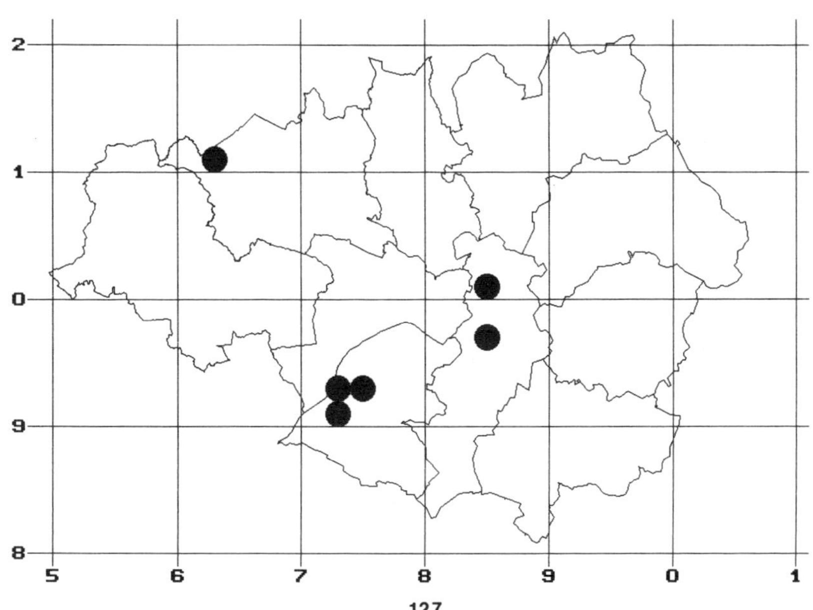